Congratulations! You have purchased a 12-month subscription to one or more *Labs On-Line,* the most intuitive, interactive, and robust electronic laboratory ... rently available. You'll be using these labs to explore biological concepts that would be difficult or impossible to cover in a standard lab.

To activate your subscription:

1. Go to http://biologylab.awlonline.com/

2. Click the "Students Register Here" button

3. Enter your pre-assigned Access Code exactly as it appears below:

Access Code: USCS-WACKO-CREPY-TIGON-ROSSI-TOUSE

4. If you purchased more than one individual Lab, and have already activated a subscription, enter your personal User ID and Password in the box where asked. We'll access your account and you can use the same personal log in for each Lab.

5. Click the Submit button

6. Complete the online registration form to activate your subscription and establish your personal User ID and Password.

7. Once your access and personal User ID and Password are confirmed, follow the Site link on the confirmation page to enter your Lab.

8. Each subsequent time you visit http://biologylab.awlonline.com, simply choose the Lab you wish to use and log in.

If you did not purchase this product new and in a shrink-wrapped package, this pre-assigned Access Code may not be valid. It can only be used once to establish a subscription, which is not transferable.

You can purchase subscriptions at your local college bookstore if your professor has specifically requested it. Subscriptions can also be purchased via credit card (MC / VISA / AMEX) by calling (800) 824-7799.

0-8053-7017-X

Student
Lab Manual for

BiologyLabs On-Line

Michael A. Palladino
Monmouth University

Robert Desharnais
Jeffrey Bell

Benjamin
Cummings

An imprint of Addison Wesley Longman

San Francisco • Boston • New York • Capetown • Hong Kong • London • Madrid
Mexico City • Montreal • Munich • Paris • Singapore • Sydney • Tokyo • Toronto

Executive Producer: Lauren Fogel
Publishing Assistant: Aaron Gass
Prepress Supervisor: Vivian McDougal
Marketing Manager: Josh Frost
Copy Editor: Jan McDearmon
Compositor: Michael Palladino
Printer: Victor Graphics, Inc.

References are made in this lab manual to Campbell, N.A., J.B. Reece, and L.G. Mitchell. *Biology*, Fifth Edition. Menlo Park, CA: Benjamin Cummings, 1999 and Campbell, N. A., Reece, J. B., *Biology*, Sixth Edition, San Francisco, CA: Benjamin Cummings 2001.

Student Manual ISBN 0-8053-7017-X

High school edition ISBN 0-8053-7068-4

2 3 4 5 6 7 8 9 10—VG—03 02 01

Benjamin Cummings
1301 Sansome Street
San Francisco, CA 94111

Table of Contents

EnzymeLab

Background

Enzymes are a diverse and important class of proteins. Biologists refer to enzymes as biological catalysts because they increase the velocity or rate of chemical reactions in living cells. You may have already used MitochondriaLab to learn about basic principles of enzyme activity, metabolic pathways, and the role of enzymes in **metabolism**. Typically, each enzyme is capable of catalyzing a reaction between a very specific molecule or set of molecules. Molecules that an enzyme reacts with are called **substrates**.

Enzymes can catalyze very specific reactions for a given substrate with incredible precision and speed, in part because of the overall shape assumed by the amino acids that constitute the enzyme. Recall that the shape of any protein is typically influenced by the **primary structure** of the protein—the linear sequence of amino acids joined together by peptide bonds to form the protein. Enzymes also exhibit **secondary** and **tertiary structure**, and those with multiple subunits require **quaternary structure** for complete activity. Remember that these arrangements of protein structure determine the overall folding and shape or conformation of the enzyme, which in turn determines its function. There are few better examples of the important relationship between protein structure and function. All enzymes have a conformation that produces an **active site**—a pocket or groove in the enzyme where the substrate binds. The active site of an enzyme is typically specific for only one substrate, because the overall three-dimensional conformation of the active site is designed to fit the molecular shape of the substrate. For certain enzymes, the amino acids forming the active site and the way these amino acids interact with and bind to a substrate have been very well characterized. This is true for **invertase**, the enzyme you will study in this lab. Details about invertase will be discussed later in this background.

When enzymes were first studied, biochemists often compared the interaction between an enzyme and its substrate to the complementary interaction between a lock and key, where the enzyme and its active site represent the lock, while the substrate represents the key that fits the lock. Although this simple analogy may help to explain the specificity of an enzyme for a substrate, in reality this interaction is much more complex. Modern-day biochemists typically acknowledge that enzyme-substrate interactions follow an *induced-fit* hypothesis. This hypothesis states that the enzyme does not merely provide a static active site into which the substrate fits. Instead, as the substrate begins to enter the active site, the shape of the active site changes—induced by the substrate as it begins to enter the active site—thus enabling the enzyme to conform around the substrate and facilitate its binding to the active site.

Substrate binding to the active site of an enzyme is only the first step toward catalyzing a reaction. If enzymes function as biological catalysts, then how can we determine and measure the efficiency and rate of any reaction that a particular enzyme is carrying out on a given substrate that it binds? To begin, it is important to remember that one key aspect of enzyme activity is that while an enzyme speeds up the rate of a reaction, it is not altered by the reaction itself. An enzyme does not become part of the reactants or products. After an enzyme has catalyzed a reaction, it releases its substrate and then the active site of the enzyme is available to bind to another fresh substrate and repeat this process. For most enzymes this process, known as the catalytic cycle of enzyme activity, can be repeated very rapidly as long as there is enough substrate for the enzyme to react upon. Regardless of the type of reaction an enzyme is catalyzing—for example, a synthesis reaction or a degradation reaction—the catalytic cycle of an enzyme is often described by the following equation:

$$E + S \rightarrow ES \rightarrow E + P$$

This cycle begins when an enzyme (E) binds to a substrate (S) to form an enzyme-substrate complex (ES). Enzyme-substrate complexes typically form as a result of weak bonds between amino acids in the active site and atoms of the substrate. Depending on the type of reaction catalyzed by the enzyme, the enzyme then manipulates the substrate into the proper conformation to catalyze a reaction—for example, breaking bonds by hydrolysis, catalyzing the formation of new bonds, or rearranging atoms in the substrate to convert the substrate into a new molecule or molecules called products (P). Once the reaction has occurred, the enzyme releases the product(s); thus, the active site is free and available to bind to another substrate so the enzyme can repeat this cycle. Enzyme-catalyzed reactions can be reversible or irreversible.

Through this cycle, enzymes are said to lower the activation energy of a reaction—the energy required to make or break chemical bonds in a substrate to initiate a reaction. It is convenient to think about activation energy as a barrier that a cell must overcome to enable a reaction to take place. To understand why this is necessary for living cells, think about the factors that can regulate the rate of most reactions that you might carry out in a test tube in a chemistry lab. Increasing the temperature of the tube as well as increasing reactant concentration in the tube are two conditions that will increase the rate of most chemical reactions. Increasing temperature raises kinetic energy of reactants in the tube, thus increasing the likelihood of collisions between molecules. Raising reactant concentration results in more frequent collisions between molecules. Both factors typically increase the rate of a chemical reaction. Although heating or cooling a test tube is an effective way to regulate enzyme activity in a controlled laboratory environment, body temperature homeostasis prevents living organisms from raising and lowering body temperature to accommodate the large number of reactions that occur simultaneously in any given cell. Similarly, substrate concentrations for many biochemical reactions in a living cell are very low—in the nanomolar to picomolar range or lower—thus, it simply is not efficient for a cell to wait for kinetic energy to cause molecules to collide randomly and react with each other.

Think of enzymes as a set of "molecular hands" for a living cell. Consider the simple analogy of a bag of separate nuts and bolts. If screwing one nut onto one bolt is the reaction you want to carry out, you could expend a lot of energy shaking the bag until, over time, one nut randomly finds its way onto the end of a bolt! The energy you expended to begin to put a nut and bolt together is activation energy. Alternatively, you could speed up this reaction by reaching into the bag and using your hands to screw a nut onto a bolt, thus reducing the amount of energy required for this reaction to occur and greatly increasing the rate of this reaction.

Although enzymes are absolutely essential for accelerating biochemical reactions, a number of conditions influence enzyme activity. Enzymes don't always operate at their maximal rate, however. Most enzymes demonstrate temperature and pH optimums—a temperature and pH at which enzyme activity is greatest. For example, as you might expect, blood enzymes perform optimally at a pH close to 7.4, the pH of normal human blood, whereas stomach enzymes have an optimal pH of around 2.0 to coincide with acidic conditions in the stomach. Varying these conditions typically affects the conformation of the enzyme, which in turn influences an enzyme's ability to bind to its substrate and catalyze a reaction. Recall that when any protein unfolds, it becomes less active or inactive; this process is called protein **denaturation**. Enzymes can become denatured in response to an increase in temperature because raising temperature can break bonds—such as **hydrogen bonds**, **Van der Waals** attractions, and disulfide bridges—that are responsible for the secondary and tertiary structure of a protein. Changes in pH can also disrupt protein structure by changing hydrogen bond and ionic bond interactions, and by changing side group (R-group) interactions that contribute to the tertiary structure of an enzyme. In the most extreme circumstances, changing temperature or pH too dramatically—for example, by boiling—can completely denature an enzyme, causing it to lose all of its activity.

When biologists study enzyme-catalyzed reactions, we are typically interested in more detailed aspects of enzyme biochemistry than just substrates and temperature and pH optimums. When studying a particular enzyme, biologists often study the interactions between the enzyme and its substrate and the reaction rate of the enzyme in great detail. One of the most common approaches for measuring enzyme kinetics is the Michaelis–Menton equation, named after Leonor Michaelis and Maud Menten, two biochemists whose landmark discoveries in the early 1900s continue to serve as the basis by which important biochemical properties of enzyme kinetics are studied.

Once factor that strongly influences enzyme kinetics is the concentration of substrate [S] available for a particular enzyme. Consider the figure shown below. If we were to plot substrate concentration against initial reaction velocity (V or V_O) as a measure of the rate of a reaction, many enzymes would show a pattern known as first-order kinetics. In first-order kinetics the rate of the reaction depends on the substrate concentration. Reaction rate increases as substrate concentration increases, because more substrate is available to be bound by the enzyme. If an enzyme were supplied with an infinite amount of substrate, then the reaction would reach a maximum velocity. This is because as the reaction proceeds, fresh substrate is rapidly binding to the active site of the enzyme, "saturating" all of the active sites for every enzyme molecule in the reaction. Under these conditions, adding additional substrate produces no effect on reaction velocity because the enzyme molecules are incapable of working any faster. This plateau of maximum velocity is abbreviated as V_{max}.

Plotting [S] versus V for an enzyme-catalyzed reaction provides us with important information on the activity of the enzyme being studied because we can use V_{max} to determine another important parameter of enzyme kinetics, the Michaelis constant (K_M), which is a measure of the affinity of an enzyme for a substrate. The Michaelis constant is equal to [S] at $-V_{max}$. To use EnzymeLab, you must be familiar with the basic factors involved in Michaelis–Menton kinetics as described in the previous two paragraphs. These factors are often expressed in the Michaelis–Menten rate equation as

$$V_O = \frac{V_{max}[S]}{K_M + [S]}$$

We can learn a great deal about enzyme activity from Michaelis–Menten measurements. In particular, K_M is a measurement of the substrate concentration required for an efficient reaction to occur. The Michaelis–Menten equation can also be used to measure k_{cat}, the turnover number, which tell us the catalytic ability of an enzyme to create product under saturated conditions. When biochemists are studying a reaction to determine if the enzyme involved follows Michaelis–Menten kinetics, data are typically plotted in one of two ways, as a Lineweaver—Burk plot or as an Eadie–Hofstee plot. These plots are created by rearranging the Michaelis–Menten equation. The Lineweaver–Burk equation is:

$$\frac{1}{V} = \frac{K_M}{V_{max}} \cdot \frac{1}{[S]} + \frac{1}{V_{max}}$$

The Eadie–Hofstee equation is:

$$V = V_{max} - K_M\frac{V}{[S]}$$

Examples of each type of plot are shown below.

In a Lineweaver–Burk plot, notice that the Michaelis equation is inverted to produce a plot with a straight line for V versus [S]. The slope of the line is K_M/V_{max}, the x-intercept tells us $-1/K_M$, and the y-intercept is $1/V_{max}$. In an Eadie–Hofstee plot, V is plotted against V/[S]. The slope of the line tells us $-K_M$, the y-intercept is V_{max}, while the x-intercept is V_{max}/K_M.

Understanding the kinetics of individual enzymes is important for understanding the overall biochemistry of living cells. However, cell metabolism is dependent on the combined actions of many different enzymes that are essential for the **anabolic** and **catabolic reactions** that must occur to maintain the physiology of a cell. It is important to understand that many reactions that require enzymes rarely involve just a single enzyme that works by an all-or-none process. This would be like trying to manufacture a car from all of its components in one sweeping motion! Instead, many biochemical reactions involve cascades of enzymatic reactions, called metabolic pathways, that serve to catalyze a series of reactions. In a metabolic pathway, several enzymes work in a sequential fashion to convert reactants into a product or products.

At each step in a metabolic pathway, intermediate molecules (metabolites) are produced that serve as the substrates for subsequent enzymes in the pathway. One benefit of a metabolic pathway compared to a single-enzyme reaction is that a cell can often precisely regulate the amount of product it can generate by independently controlling the catalytic activity of certain enzymes in the pathway.

Intermediate molecules are often an important part of the control of a metabolic pathway. One way in which metabolic pathways can be regulated by intermediates involves **feedback inhibition**. In this regulatory process, a final or end-product of a reaction can inhibit enzymes in the metabolic pathway. This process allows a cell to carefully control the amount of end-product that it produces as a way to prevent excess accumulation and waste of an end-product. Frequently, the end-product inhibits or blocks the activity of an enzyme at one of the initial, rate-limiting steps in the pathway to prevent the unnecessary production of intermediates. This would be similar to a car manufacturer, whose car supply has exceeded the public's demand, stopping the auto assembly line at the first step rather than halfway through the assembly process to avoid producing half-completed cars.

For the purpose of studying cell metabolism, as well as for medical treatment and other applications, we can use molecules as enzyme inhibitors to artificially regulate enzyme activity. Some of these inhibitors can bind to enzymes in a reversible or an irreversible fashion. The two primary classes of enzyme inhibitors are called competitive and noncompetitive inhibitors. **Competitive inhibitors** are molecules that decrease an enzyme's activity by binding to the active site of the enzyme and preventing binding of the enzyme's normal substrate. Competitive inhibitors can work in this fashion because their molecular shape so closely resembles the natural substrate that the inhibitor is able to "compete" with the substrate for binding to the enzyme's active site. Although competitive inhibitor molecules bind to the active site of an enzyme, catalysis does not occur.

Noncompetitive inhibitors also reduce or block an enzyme's activity, but unlike competitive inhibitors these molecules do so by binding to a portion of the protein other than the active site. This binding results in a change in the overall three-dimensional conformation of the protein—altering the active site so that it will not bind to the substrate.

Because many enzymes are subject to inhibition by inhibitor molecules, biologists can take advantage of this property by designing compounds that can be used to modify enzyme activity for the purpose of treating human ailments and disease, as well as a number of other applications. For example, one of the more widely known competitive inhibitors is the antibiotic penicillin, which functions to block the active site of an enzyme required for cell wall synthesis by bacteria. Another example of a commonly used inhibitor is aspirin, which functions as a noncompetitive inhibitor of an enzyme involved in the production of prostaglandins, molecules that cause fever and inflammation. Last, two drugs employed in the treatment of acquired immunodeficiency syndrome (AIDS) function to inhibit enzymes in the human immunodeficiency virus (HIV). The first anti-HIV drug developed for patient use, AZT (3'-azido-2,'3'-dideoxythymidine), competitively inhibits an enzyme required for the synthesis of HIV DNA. Another class of enzymes, called protease inhibitors, function to reduce replication of HIV by competitive inhibition of an essential viral enzyme called HIV protease.

Now that you are familiar with some important biochemical properties of enzymes, it is time to put your knowledge to work. Countless numbers of different enzymes exhibit many of the properties discussed in this background. In EnzymeLab you will work with an enzyme that most likely played an important role in digesting some of the food molecules that you ate this morning for breakfast! The enzyme chosen for this lab is **invertase** (β-fructofuranosidase; E.C. 3.2.1.26), also commonly called sucrase and saccharase. This enzyme uses water to catalyze the **hydrolysis** of the disaccharide sucrose. Sucrose, or "table sugar," is technically called α-D-glucopyranosyl $(1\rightarrow2)$ β-D-fructofuranoside because it consists of the two simple sugars glucose (α-D-glucopyranosyl) and fructose (β-D-fructofuranoside) joined together by a glycosidic bond connecting carbon 1 of glucose to carbon 2 of the fructose molecule. Invertase cleaves the glycosidic bond between glucose and fructose. In animals, this reaction is required to digest fructose and release the monosaccharides glucose and fructose to make them readily available for absorption into the bloodstream. This reaction is important because most organisms cannot metabolize sucrose directly—it must first be converted to monosaccharides for cells to utilize sucrose as an energy source.

Invertase is present in a wide range of organisms including animals, plants, yeast, fungi, and algae. In humans, invertase is found on the surface of epithelial cells lining the inner walls of the small intestine. Depending on the source of invertase, the enzyme is active at a range of temperatures from 40°C to 70°C. Invertase can also be active in the pH range from 4.0 to 10.0. Two of your goals for this laboratory are to determine the optimal temperature and pH for the invertase you will study in EnzymeLab.

You will use EnzymeLab to study important biochemical parameters of invertase. You will set up an experiment by adding substrate to a test tube along with purified enzyme to determine temperature and pH optimums for invertase, to measure and calculate values of Michaelis–Menten kinetics, and to study the effect of inhibitors on invertase activity. Studying invertase will help you understand how biochemists determine the specific biochemical properties of an enzyme that follows Michaelis–Menten kinetics.

References

1. Mathews, C. K., van Holde, K. E., and Ahern, K. G. *Biochemistry*, 3rd ed. Menlo Park, CA: Benjamin/Cummings, 2000.

2. Nelson, D. L., and Cox, M. M. *Lehninger Principles of Biochemistry*, 3rd ed. New York: Worth, 2000.

3. Schomburg, D., and Stephan, B., eds. *Enzyme Handbook*. Heidelberg, Germany: Springer-Verlag/Berlin, 1999.

Introduction

In this laboratory, you will perform simulations of experiments designed to study the biochemistry of the enzyme invertase, an important enzyme involved in the metabolism of the disaccharide sucrose. You will learn to measure and calculate important parameters of enzyme kinetics, and to measure product formation by spectrophotometry. Also, by changing reaction conditions and plotting enzyme kinetic data, you will learn about factors that influence the catalytic activity of invertase.

Objectives

The purpose of this laboratory is to:

- Help you understand important properties of enzyme-catalyzed reactions, and techniques that are used to measure enzymatic reactions.
- Demonstrate how changes in important parameters of enzyme biochemistry, such as substrate concentration, can affect enzyme activity.
- Study and measure Michaelis–Menton enzyme kinetics.
- Investigate pH and temperature optimums of invertase.
- Examine the effects of different classes of enzyme inhibitors on invertase.
- Simulate the spectrophotometric measurement of enzyme-catalyzed reactions.

Before You Begin: Prerequisites

Before beginning EnzymeLab you should be familiar with the following concepts:

- Basic principles of thermodynamics, metabolism, and metabolic pathways (see Campbell, N. A., Reece, J. B., and Mitchell, L. G. *Biology* 5/e, chapter 6).
- The structure of sucrose and its importance as a source of energy (chapter 5).
- Properties and functions of protein enzymes as biological catalysts (chapter 6).
- Temperature and pH as important factors that affect enzyme activity (chapter 6).
- Feedback mechanisms involved in the control of enzyme activity and metabolism, and different types of enzyme inhibitors including competitive inhibitors and noncompetitive inhibitors (chapter 6).
- First-order enzyme kinetics measurements of the Michaelis–Menten equation including V_O, V_{max}, [S], and K_M.
- Analysis of kinetic data by the following plots: V vs. [S], V vs. pH, V vs. temperature, Lineweaver–Burk, and Eadie–Hofstee plots.
- Basic operations of a visible light spectrophotometer and its application in measuring enzyme kinetics.

Assignments

For your ease in completing each assignment, the background text relevant to the experiment that you will perform is *italicized*, instructions for each assignment are indicated by plain text, and questions or activities that you will be asked to provide answers for are indicated by **bold text**.

The following assignment is designed to help you become familiar with the operation of EnzymeLab.

Assignment 1: Getting to Know EnzymeLab: Setting Up an Experiment

The first screen that appears in EnzymeLab shows you a biochemistry lab containing all the reagents and equipment you will need to perform your experiments.

Click on each item in the lab to learn more about its purpose. Once you are familiar with the lab, click on the Experiment button to begin the first assignment. This assignment is designed to help you become familiar with the operation of EnzymeLab.

When you set up an experiment in EnzymeLab, you will add a buffered solution, fructose as substrate, invertase, and, in some reactions, inhibitors to a test tube to measure the rate of invertase activity. You will have the choice of performing each reaction at different temperatures and under different buffer conditions in which you can examine the effect of these parameters on invertase activity. A visible light spectrophotometer will measure product formation by measuring the absorbance of glucose (released as fructose is cleaved by invertase) at approximately 450 nm. Data are recorded and plotted as a function of product concentration [P] in micromoles (μm) versus time (minutes). Raw data you collect can then be analyzed by several different types of plots that are commonly used for analyzing kinetic data for enzyme-catalyzed reactions.

1. Effect of Temperature on Invertase Activity
    Changes in temperature can dramatically influence the activity of most enzymes by affecting enzyme structure. This exercise is designed to help you learn how to set up an experiment in EnzymeLab and understand the effect of temperature on enzyme activity. You will also analyze data from this experiment to determine the ideal temperature optimum for invertase activity.

    To begin any experiment, you first need to set the temperature of your water bath, then add buffer and substrate to the reaction tube. As is the case with real experiments in a biochemistry lab, enzyme should always be added to the tube last to prevent the reaction from starting before all necessary components have been added to your test tube.

    Develop a hypothesis to predict the effect of an increase in temperature on invertase activity, then test your hypothesis as follows.

    Notice that the default temperature for the water bath is 40°C. To change the temperature, you can either enter a temperature value in the text box or use the arrows. Change temperature to 30°C; this is the lowest temperature at which you can carry out an experiment. Notice that the default buffer pH is 7.0. Do not change the default pH for any measurements in this experiment.

    Notice that the default value for substrate concentration [S] is indicated by the slider bar labeled [S] and the value of 25 mM appears in the text box to the far right of the [S] slider. Substrate concentration for sucrose is reported in millimolar (mM) units. You can change [S] by either moving the slider bar or typing a value in the [S] text box. For the first reaction in this experiment we will begin by carrying out a reaction with 90 mM sucrose. Change [S] to 90 mM. Do not select any inhibitors for this experiment.

Note: The [S] value that you use for any experiment will be reported in the table under the Plot Data view. However, two students running an experiment with the same [S] may see slightly different results because EnzymeLab is programmed to simulate differences that represent slight experimental variations in concentration measurements such as you would encounter if setting up this experiment in a wet lab.

To add enzyme to your reaction tube, click the Add Enzyme & Go! button. This will also activate the spectrophotometer to measure product concentration.

a. Determining Starting Velocity (V_O)
After each time you add enzyme, a plot of product concentration versus time will appear with enzyme kinetic data plotted as data points in solid black circles.

What did you observe for the plot of product concentration versus time? Is this what you expected? Explain your answer.

To begin your analysis of this experiment, you first need to determine the starting velocity of the reaction (V_O). This is easily accomplished because V_O (the initial rate of the reaction showing first-order kinetics where the rate of the reaction depends on [S]) is represented by the slope of the linear portion of the curve in this type of plot. The plateau (asymptote) of the plot represents zero-order kinetics where the rate of the reaction does not depend on [S]. Note: For many of the reactions you will run, the reaction will not reach zero-order kinetics.

To determine the slope of the line, a red line will appear on the plot. You can then click on this red line and move it to find the best fit for the slope of the plotted points for invertase activity. Try to align the red line so that you have an equal number of data points bisected by the line and an equal number of points above and below the line. This is generally a good approach for finding the best-fit slope. The program will not tell you when you have found the best fit, so use your best judgment! Follow the directions below to determine the slope of the line for this first measurement.

Click within the plot and move the red line until you have found the slope of the plotted data points. Before we can analyze this information further, you must record your data by clicking the Record Data button. Note: The Record Data feature is not active until you have properly determined the slope of the plotted line.

After you have recorded data for a measurement, you must click the Clear Experiment button before you can take another measurement. Note: If you forget to record data and attempt to clear the experiment, a warning box will appear asking if you want to clear data without recording it.

Keeping buffer pH constant and [S] at 90 mM, create another experiment but this time increase temperature to 35°C. Run the experiment, determine the slope of the line, then record this measurement. Repeat this process to set up experiments increasing temperature in 5-degree increments (40°C, 45°C, 50° etc.) until you reach the maximum temperature of 85°C. Find the slope of the line for each experiment and record these data.

Click on the Plot Data button to prepare a plot of your data for this experiment.

b. Plotting Invertase Kinetic Data
In the first window that appears in the Plot Data view you can title the plots for the experiment you are working on, select a plot to create, change the symbols for this plot, and view raw data in tabular form showing each measurement for a particular experiment—[S], the presence or absence of an inhibitor, inhibitor concentration [I], temperature, pH, and V_O.

From the Plot Data view we can now use EnzymeLab to carry out a number of important calculations and present these data as plots that are traditionally used for studying enzyme kinetics. Each of the plots available in EnzymeLab is briefly described below. It is very important that you are comfortable with the abbreviations referred to below and the purpose of plotting data with these different types of plots. These principles form the basis for the majority of experiments that you will be carrying out with EnzymeLab.

V_O vs. [S]: convenient way to express the relationship between reaction velocity (Vo) and substrate concentration [S]. Used to determine V_{max}, the maximum velocity of an enzyme, represented by the asymptote (plateau) of the line. We can also use this plot to measure the Michaelis constant (K_M). To find the Michaelis constant, you need to locate -V_{max} then find where this value would intersect the x-axis. The [S] represented by this intersect point is the K_M. This type of plot is a simple way to represent and determine V_{max} and K_M; however, it is not the most accurate way to determine these values because we are extrapolating from the plot.

V_O vs. Temperature: analyzes the effect of temperature on reaction velocity.

V_O vs. pH: analyzes the effect of pH on reaction velocity.

Lineweaver–Burk: also called a double-reciprocal plot. Produces a linear plot for the inverse of velocity (1/V) versus the inverse of substrate concentration [1/S]. Used to determine two important characteristics of enzymes that follow Michaelis–Menton kinetics: K_M (substrate concentration at which a reaction has reached half of its maximum velocity), and k_{cat} (turnover number = number of substrate molecules undergoing a reaction per enzyme molecule per second). The Lineweaver–Burk plot is a more accurate plot for determining V_{max} and K_M than a plot of V_O versus [S] because Lineweaver–Burk plots are based on algebraically arranged equations of the Michaelis–Menten equation.

Eadie–Hofstee: a plot of V versus V/[S]. Another way to determine parameters of Michaelis–Menton kinetics. The y-intercept indicates V_{max}, the x-intercept determines V_{max}/K_M, while the slope of the line determines $-K_M$.

Note: **Any of the plots that you generate can be saved to disk or printed by clicking on the Export Graph button, which appears to the left of each plot. Clicking on this button will open a separate window with your plot. From this window you can then save your plot to your hard drive or a disk, and you can print your plot by using the print feature of your browser software.**

Plotting V_O vs. Temperature: Click in the Title box and type in the title "Experiment 1 – Temperature Optimum." V_O vs. [S] should appear as the default plot in the Plot Type box. Click on the popup menu and select V_O vs. Temperature. For any plot, you must first select the data that you want to plot. You can click on an individual row to select it and the row will be highlighted, or you can select several rows by holding down the Shift key and clicking on each row. Shift-click on each row of measurements that you recorded for this temperature experiment, then click the Plot Selected Data button to produce a plot of V_O versus temperature.

Click anywhere on the V_O vs. temperature plot and drag the vertical gray dashed line to locate the highest value for enzyme activity (V_{max}). When you have correctly located this value, the gray line will become black and it will freeze in place. This value, indicated in the best-temperature text box, represents the optimal temperature for invertase activity under these conditions of pH and [S].

What is the optimal temperature for invertase activity? Is this what you expected? Would invertase isolated from any two organisms (for example, yeast invertase vs. invertase from the small intestine of humans) show the same temperature optimum? Why or why not? Explain your answers.

Explain why temperatures lower or higher than the optimum cause decreases in invertase activity. What is happening to the enzyme to produce these decreases in activity?

Carefully examine the curve for V_O vs. temperature. Is the slope of the line on <u>both sides</u> of the curve the same or different? If the slope of the line to the left of maximum velocity is different from the slope of the line to the right of maximum velocity, explain why this is. What is responsible for these differences in enzyme kinetics?

If you were to carry out these temperature experiments at a higher or lower [S], what effect would [S] have on the temperature optimum for invertase? Formulate a hypothesis and then test your hypothesis. What did you discover? Explain your results.

If you were to carry out these temperature experiments at a higher or lower pH value, what effect would this have on the temperature optimum for invertase? Formulate a hypothesis and then test your hypothesis. What did you discover? Explain your results.

Assignment 2: pH Optimum for Invertase

Another factor that strongly influences enzyme activity in living cells is the pH of the environment in which the enzyme is designed to function. For example, in humans, a cytoplasmic protein in a skin cell is surrounded by a different fluid environment at a different pH than a membrane-bound enzyme like invertase with an active site that projects out into the lumen of the small intestine. The following exercise is designed to help you understand the effect of pH on enzyme activity by studying invertase activity over a range of different pH values from acidic to basic conditions.

Formulate a hypothesis to predict the effect of pH on invertase activity.

Set up an experiment at the optimal temperature that you determined in assignment 1, with a substrate concentration of 90 mM. Beginning at the lowest pH value, 3.0, measure invertase activity, find the slope of the line, record data, and repeat this process for other buffer solutions with different pH values. Run experiments for at least two different buffers for each whole number change in pH units until you reach the maximum pH value of 10.0 (e.g., 3.0, 3.4, 4.0, 4.4, 5.0, 5.4).

Create a plot of V_O vs. pH. Click on this plot and drag the gray dashed pH line until you find V_{max}. When this line is correctly aligned, you will have found the pH optimum for invertase under these reaction conditions. The optimal pH value will appear in the best pH text box.

What is the optimal pH for invertase activity? Do the results of this experiment support or refute your hypothesis? Why or why not?

Explain your answers? Why and how do pH changes affect invertase activity?

If you were to carry out these pH experiments at a higher or lower temperature, what effect would this have on the pH optimum for invertase? Formulate a hypothesis and then test your hypothesis. What did you discover? Explain your results?

Assignment 3: The Effect of Substrate Concentration on Invertase Activity

This exercise is designed to help you interpret Michaelis–Menten parameters for invertase by learning how to create and interpret plots of enzyme kinetics. In this experiment, you will add different concentrations of sucrose to invertase and then study the effects of sucrose concentration on invertase activity.

Set up an experiment at 50°C with a buffer pH of 4.0 and a [S] of 0.0 mM. Add enzyme. Find the slope of the line and record data for this first measurement. Continue to take measurements by carrying out separate experiments with increasing concentrations of [S]. Take a measurement at 2.5 mM, 5 mM, and 10 mM, then continue in 10 mM increments (e.g., 20 mM, 30 mM, 40 mM, 50 mM, 60 mM) until you have reached a [S] of 400 mM. Find the slope and record data after each measurement. Once you have finished these measurements, click the Plot Data tab.

Follow the directions below to use these data to create some of the plots described in assignment 1. Each plot is selected from the popup menu for Plot Type. Refer to these directions if necessary when creating plots for other experiments that you are working on.

Plotting V_O vs. [S]: Click in the Title box and type in the title "Experiment 1 – Substrate Concentration." V_O vs. [S] should appear as the default plot in the Plot Type box. Shift-click on each row of measurements from 0 to 100, then click the Plot Selected Data button to produce a plot of V_O versus [S].

Before you can determine V_{max} and K_M from this plot, you must first find the best-fit line for these data points. Click on the black K_M arrow at the x-axis. Slide this arrow until you have found the best-fit line for these points. Adjust this line as needed by clicking and sliding the black V_{max} arrow at the y-axis. A numerical value for V_{max} and K_M will appear in the V_{max} and K_M text boxes.

Setting Up a Lineweaver–Burk Plot: Return to the Data view and plot your data as a Lineweaver–Burk plot. Find the best-fit line for these data points by clicking and sliding the arrows at the x-axis and y-axis. The arrow at the intersect of the black plotted line and the y-axis indicates $1/V_{max}$. Notice that a value for $1/V_{max}$ will appear in the $1/V_{max}$ text box beneath the curve box. The arrow at the intersect of the plotted line and the x-axis determines $-1/[S]$.

For a Lineweaver–Burk plot, V_{max} and K_M will not be calculated for you; you must calculate these values yourself. Based on the values you determined for $1/V_{max}$ and $-1/[S]$ calculate V_{max} and K_M from your plot. For your convenience a calculator can be accessed by clicking the Calculator tab at the top of the screen.

Setting Up an Eadie–Hofstee Plot: Return to the Data view and plot your data as an Eadie–Hofstee plot. Find the best-fit line for these data points by clicking and sliding the arrows at the x-axis and y-axis. The black arrow at the intersect of the black plotted line and the y-axis indicates V_{max}. Notice that a value for V_{max} will appear in the V_{max} text box. The arrow at the intersect of the plotted line and the x-axis indicates V_{max}/K_M.

For an Eadie–Hofstee plot, K_M will not be calculated for you. Calculate K_M from your plot. For your convenience, a calculator can be accessed by clicking the Calculator tab at the top of the screen.

What was the relationship between [S] and invertase activity? Was this relationship what you expected? Why or why not? Explain your answers. What was the V_{max} and K_M for invertase for this experiment? What do these values tell you about invertase and its affinity for sucrose as its substrate?

How did the values for V_{max} and K_M derived from each of the three plots compare? Were these values similar or different? If they were different, explain possible reasons for these differences. Which plot did you find easier to use for determining these values?

What do you think would happen if you carried out an experiment with concentrations of sucrose greater than 400 mM? Formulate a hypothesis, then carry out an experiment to test your hypothesis. What did you observe? Describe your results.

Assignment 4: Effect of Inhibitors on Invertase Activity

Enzyme inhibitors (I) have played an important role in helping biochemists understand how enzymes function. By inhibiting enzyme activity, it is possible to learn a great deal about the biochemical properties of a particular enzyme. For example, in the absence of substrate, competitive inhibitors can be studied according to the following reaction:

$$E + I \rightarrow EI \rightarrow E + P$$

The equilibrium or dissociation constant for inhibitor binding (K_I) in enzyme-inhibitor complexes is written as

$$K_I = \frac{[E][I]}{[EI]}$$

Typically, when determining how an inhibitor functions, experiments are carried out by measuring enzyme activity in the presence of the inhibitor and different concentrations of substrate. By comparing kinetic data for the inhibition studies to data from uninhibited reactions, it is possible to distinguish competitive inhibitors from noncompetitive inhibitors. When data from such studies are plotted in a Lineweaver–Burk plot, the slope of the plotted line indicates the apparent K_M in the presence of inhibitor (αK_M). From this value, K_I can be determined from the expression

$$\alpha = 1 + \frac{[I]}{[K_I]}$$

Two of the invertase inhibitors included in this lab are acarbose and Discorea rotundata invertase inhibitor B (DRI inhibitor B). The following assignments are designed to help you distinguish between different types of inhibitors by analyzing the effects of each inhibitor on invertase activity. Your goal is to determine how each inhibitor affects invertase.

1. Role of Acarbose as an Invertase Inhibitor
    This experiment is designed to help you determine whether acarbose functions as a competitive or noncompetitive inhibitor of invertase.

    Set up an experiment at 50°C, buffer pH 4.0, at a [S] of 50 mM with no acarbose. Run the experiment, determine the slope of the line, record data, and then repeat this type of experiment increasing [S] concentration (keeping [I] at 0.0 µM) in 10 mM increments until you have run experiments at 60 mM, 70 mM, 80 mM, and 90 mM. These uninhibited measurements will be important for studying the activity of invertase when it is inhibited by acarbose.

    Repeat this series of measurements in the presence of 0.2 µM acarbose. Keep all other conditions the same as you did for the uninhibited measurements.

    Plot a Lineweaver–Burk plot and/or an Eadie–Hofstee plot with <u>both</u> the uninhibited and inhibited data on the same plot as follows. Shift-click to select the five uninhibited measurements, select the plot you want (either Lineweaver–Burk or Eadie–Hofstee), then click Plot Selected Data. Return to the Data view by clicking the Data tab at the top of the screen. Shift-click to select the five inhibited measurements, then change the Data for Curve value to 2 (this indicates that the inhibited data will be plotted as the second curve on plot 1). Change the shape of the symbol to be plotted for these data by clicking on the popup menu for Symbol and choosing a different symbol for the inhibited data. Change the color of the inhibited data from the default, black, to another color using the Color popup menu, then click Plot Selected Data. You will now see a plot with two plotted lines. Your uninhibited data will be plotted in black and your inhibited data will be plotted in the color that you chose. Print this plot.

Determine V_{max} and K_M for the uninhibited and inhibited studies, then answer the following questions.

Compare your data from the inhibited reactions to your data from the uninhibited experiments. What did you find? Explain what happened to invertase activity as you increased [S]. Why did this occur? What happened to V_{max} in the presence of the inhibitor? What happened to K_M? If either V_{max} or K_M changed, explain why.

Based on these results and what you already know about inhibitors of enzyme activity, is this inhibitor functioning as a competitive inhibitor or a noncompetitive inhibitor? How do you know? Explain your answers.

2. Role of DRI Inhibitor B as an Invertase Inhibitor
    To investigate how DRI inhibitor B inhibits invertase, carry out the same sets of experiments, both uninhibited and inhibited, with DRI inhibitor B that you set up for acarbose. For inhibited measurements, set DRI inhibitor B to 25 mM. Note: If you did not delete the uninhibited measurements from your experiment with acarbose, you do not need to repeat the uninhibited experiments; you can use your data from the acarbose experiment.

    Compare your data from the inhibited reactions to your data from the uninhibited experiments. What did you find? Explain what happened to invertase activity as you increased [S]. Why did this occur? What happened to V_{max} in the presence of the inhibitor? What happened to K_M? If either V_{max} or K_M changed, explain why.

    Based on these results and what you already know about inhibitors of enzyme activity, is this inhibitor functioning as a competitive inhibitor or a noncompetitive inhibitor? How do you know? Explain your answers.

    Compare and contrast what happened as you increased [S] with acarbose to what happened with DRI inhibitor B. Were the results the same or different? If the effect of [S] on invertase was different in the presence of each inhibitor, explain why.

Assignment 9: Group Assignment

In the previous assignment, you learned about the effect of acarbose and DRI inhibitor B on invertase activity. A third inhibitor, l-arabinose, also inhibits invertase. Your goal is to work together in a group to determine the mechanism by which l-arabinose inhibits invertase. Work together in a group of four students to complete these exercises.

Divide your group into pairs. For one pair of students, set up and design a set of <u>controlled</u> experiments to measure the K_I for l-arabinose, then carry out these experiments. This experiment should be carried out with different concentrations of substrate and inhibitor. Plot data from each measurement as separate Lineweaver–Burk plots. Determine V_{max} and K_M for each measurement. The K_M in the presence of each different concentration of l-arabinose is the apparent K_M (αK_M). On graph paper, set up a Lineweaver–Burk plot of αK_M versus [I]. If l-arabinose is acting as a competitive inhibitor, then K_I is represented by the value of [I] that results in a K_M of twice the normal K_M.

If K_M is unaffected by l-arabinose, use graph paper to plot $1/V_{max}$ versus [I]. V_{max} in the presence of inhibitor is the apparent V_{max}. Estimate the value of [I] that reduces V_{max} by approximately one-half. This value represents K_I.

For the second pair of students, experiment. Based on your previous assignments, carry out uninhibited and inhibited experiments at a range of different substrate concentrations under conditions of optimal pH and temperature. Choose a [S] that is approximately 2–5 times higher than K_M and arbitrarily choose a concentration of l-arabinose for these studies. Run two additional sets of experiments with the same conditions, but only run one set of experiments with a higher [I] and another set with a lower [I]. Plot data as needed to evaluate your results. When possible, plot both inhibited and uninhibited data on the same plots.

Once each pair of students has evaluated their data, discuss and answer the following questions

What is the K_I for l-arabinose? Was K_M affected by l-arabinose? Was V_{max} affected by l-arabinose? What do the plots of apparent K_M and apparent V_{max} tell you about what type of inhibitor l-arabinose is?

Compare the inhibition of invertase by l-arabinose with the results obtained from inhibiting invertase with acarbose and with DRI inhibitor. Were the results similar or different? How were these results similar? How did these results differ? Explain your answers by describing what happened to V_{max} and K_M in the presence of l-arabinose.

Based on your results and your knowledge of enzyme inhibitors, does l-arabinose appear to be acting as a competitive inhibitor, noncompetitive inhibitor, or by another mechanism? Consult a biochemistry text for more details if necessary. Once you have a theory on how l-arabinose is working to inhibit invertase, describe your theory and explain how your theory is supported by your data. Design and run additional experiments if necessary until you have generated sufficient data to propose a theory for how invertase is inhibited by l-arabinose. Discuss your experiments and results with your instructor to help you answer this question.

MitochondriaLab

Background

Enzymes are an important class of proteins that function as catalysts in living cells. You may have already used EnzymeLab to learn about basic principles of enzyme activity and the role of enzymes in **metabolism**. Cell metabolism is dependent on the combined actions of many different enzymes that are essential for the anabolic and catabolic reactions that must occur to maintain the physiology of a cell. It is important to understand that many reactions that require enzymes rarely involve just a single enzyme that works by an all-or-none process. This would be like trying to manufacture a car from all of its components in one sweeping motion! Instead, many biochemical reactions involve cascades of enzymatic reactions called metabolic pathways. In a metabolic pathway, several enzymes work in a sequential fashion to convert reactants into a product or products. At each step in a metabolic pathway, intermediate molecules (metabolites) are produced that serve as the **substrates** for subsequent enzymes in the pathway. One benefit of a metabolic pathway compared with a single-enzyme reaction is that a cell can often precisely regulate the amount of product generated by independently controlling the catalytic activity of certain enzymes in the pathway.

Intermediate molecules are often an important part of the control of a metabolic pathway. One way in which metabolic pathways can be regulated by intermediates involves a process called **allosteric regulation**. In this form of control, a molecule binds to a portion of an enzyme, other than the active site, and alters the shape of the active site to either stimulate or inhibit the enzyme. **Feedback inhibition** is another mechanism that cells can use to regulate a metabolic pathway. In feedback inhibition, a final or end-product of a reaction can inhibit enzymes in the metabolic pathway. This process allows a cell to carefully control the amount of end-product that it produces as a way to prevent excess accumulation and waste of an end-product. Frequently, the end-product inhibits or blocks the activity of an enzyme at one of the initial, rate-limiting steps in the pathway to prevent the unnecessary production of intermediates. This would be similar to a car manufacturer, whose car supply has exceeded the public's demand, stopping the auto assembly line at the first step rather than halfway through the assembly process to avoid producing half-completed cars.

One of the most essential metabolic pathways that occur in all living cells involves the enzymatic conversion of food molecules to produce energy in the form of a molecule called **adenosine triphosphate (ATP)**. ATP is a critical energy source universally used by all living cells. ATP powers many chemical reactions by providing the phosphate groups necessary for phosphorylation reactions. Phosphorylation reactions are catalyzed by a broad class of enzymes called kinases. Kinases can transfer a phosphate group from ATP onto other molecules. Phosphorylation of molecules such as proteins often results in a change in the shape of the protein, which activates the protein to perform a desired function. For example, muscle contraction requires that the contractile proteins in a muscle cell—actin and myosin—be phosphorylated, thus enabling them to slide along one another to produce shortening of the muscle cell. Although lipids and proteins can undergo catabolism in cells to serve as metabolic fuels to power the reactions necessary for ATP synthesis, carbohydrates such as glucose are excellent sources of energy for producing ATP. Many plant and animal cells, including human cells, can convert glucose into ATP in the presence of oxygen by a set of reactions called **aerobic cellular respiration**. Other cells—for example, certain yeast cells, bacteria, and human skeletal muscle cells—can produce ATP from glucose in the absence of oxygen (**anaerobic** conditions) via reactions called **fermentation**. Cellular respiration produces a much higher total yield of ATP from one molecule of glucose than does fermentation, while producing a minimal number of waste products. In addition to ATP, other end-products produced include water and carbon dioxide, the primary waste product produced by cellular respiration. The summary equation showing the net yield of products created by the catabolism of one molecule of glucose ($C_6H_{12}O_6$) is shown below:

$$C_6H_{12}O_6 + 6\,O_2 \rightarrow 6\,H_2O + 6\,CO_2 + 36\,ATP$$

Glucose catabolism via cellular respiration can be grouped into three major metabolic stages: (1) **glycolysis**; (2) the **Krebs cycle**, also known as the citric acid cycle or tricarboxylic acid cycle (TCA cycle); and (3) the **electron transport chain** and **oxidative phosphorylation**. In eukaryotic cells, glycolysis occurs in the cytoplasm of the cell. The Krebs cycle occurs in the **mitochondrial matrix**, while the reactions of the electron transport chain and oxidative phosphorylation occur on the **cristae** of the mitochondrion. These pathways rely on oxidation-reduction reactions in which electrons are enzymatically removed (**oxidation**) from glucose and transferred (**reduction**) to electron acceptor molecules such as **nicotinamide adenine dinucleotide (NAD⁺)**. Upon receiving electrons, NAD⁺ is reduced to NADH, which functions as an electron carrier that supplies electrons to an electron transport chain in mitochondria that will ultimately power ATP synthesis in the reactions known as oxidative phosphorylation. The reactions of cellular respiration are summarized in the next three paragraphs.

Glycolysis involves ten enzymatic steps that ultimately result in the degradation of one molecule of glucose into two molecules of a three-carbon acid called pyruvate (pyruvic acid). Glycolysis also results in the net production of two ATP molecules and two molecules of NADH. These reactions can occur under aerobic or anaerobic conditions, but the yield of pyruvate and NADH is always the same. Before entering the Krebs cycle, each pyruvate is subsequently converted into a molecule called acetyl coenzyme A (**acetyl CoA**).

The Krebs cycle involves eight enzymatic steps that serve to oxidize both molecules of acetyl CoA. The primary result is the production of six molecules of NADH and two molecules of an electron carrier molecule similar to NADH called flavin adenine dinucleotide ($FADH_2$). These reactions were originally known as the tricarboxylic acid cycle or citric acid cycle because each enzymatic step results in the production of eight organic acids called carboxylic acids (citrate, isocitrate, α-ketoglutarate, succinyl CoA, succinate, fumarate, malate, and oxaloacetate) as intermediate molecules. Each of these molecules serves as a substrate for the next enzyme in the Krebs cycle, and some of these intermediates serve to regulate enzymes of cellular respiration. You will design experiments with some of these intermediates using MitochondriaLab to test the effects of each intermediate on the reactions of cellular respiration.

During the reactions of the electron transport chain, the electrons stored in the NADH and $FADH_2$ produced by the Krebs cycle are enzymatically transferred to a series of electron acceptor molecules in the cristae. Many of the electron acceptor molecules are iron-containing proteins called **cytochromes**. The cytochromes are clustered together within the cristae to form four multiprotein groups, or complexes of proteins—named groups I, II, III, and IV—that transfer electrons in a sequential fashion. Molecular oxygen serves as the final electron acceptor in this chain. Oxygen molecules receive electrons from complex IV. This reaction represents the oxygen-consuming stage of cellular respiration because reduced oxygen molecules are converted into water during this transfer. As electrons are transferred along this chain, this series of oxidation-reduction reactions generates a hydrogen ion (H^+) gradient (called a **proton-motive force**) in the intermembrane space of the mitochondrion. This H^+ gradient provides the energy necessary for an enzyme called **ATP synthase**. This enzyme also functions as an ion channel to allow H^+ flow down a gradient from the intermembrane space into the mitochondrial matrix. The H^+ flow through ATP synthase activates the enzyme to synthesize ATP from ADP and inorganic phosphate in a final stage called oxidative phosphorylation.

Studying the reactions of cellular respiration is an excellent way to develop an understanding of the complexities of cell metabolism and the mechanisms by which a cell can control its metabolism. One way to study the reactions of cellular respiration involves performing experiments on preparations of mitochondria that have been isolated from mitochondria-rich tissue, such as skeletal muscle or liver, using cell fractionation techniques. Many of the details of these reactions have been well understood for almost a hundred years. Numerous scientists were involved in carrying out the biochemical experiments that were used to learn about the substrates, enzymes, and cofactors required for cellular respiration, the sequence of each reaction relative to one another, and the regulation of these reactions. These reactions were some of the first well-characterized and well-understood examples of a metabolic pathway.

Some of the landmark experiments that led to our understanding of these reactions involved classic techniques in biochemistry that employed glass reaction flasks into which diced tissue preparations, intermediates, end-products, waste products, and inhibitors could be added independently, in combination, or in sequence. Biochemical activities such as the production of end-products, consumption of intermediates, and rates of enzyme activity were then measured to follow the rates of the reaction. Modern-day biochemists continue to rely on experiments of this design to study metabolism.

Many investigators contributed to research in the late 1800s and early 1900s that unraveled the sequence of reactions that occur during glycolysis. Biochemical studies in the early 1900s demonstrated that preparations of diced animal tissues exhibited the enzymatic ability to carry out oxidation-reduction reactions by transferring hydrogen ions from acids such as citrate, malate, and fumarate to dye molecules that would change color when they were reduced. Albert Szent-Gyorgi carried out some of the earliest studies. When performed under aerobic conditions, his studies established that organic acids could be oxidized to produce carbon dioxide and water, and that inhibitors of these reactions blocked the consumption of oxygen in the reaction.

Another pioneer of cellular respiration was the German biochemist Hans Krebs, who discovered the citric acid cycle, which now bears his name—the Krebs cycle. Krebs proposed that this pathway involved a cycle of enzymatic reactions that were required to oxidize organic acids such as pyruvate. Many of Krebs's experiments involved studying the oxidation of organic acids in slices of flight muscle from pigeons. To support active wing flapping over long distances, these muscle cells depend heavily on oxidative metabolism to produce ATP, so they served as an excellent tissue source for Krebs's research. Krebs also determined that the use of inhibitors such as malonate could block the production of some intermediates in the cycle while causing the accumulation of other intermediates. Using this approach, Krebs was able to determine the sequence of enzyme steps involved in the citric acid cycle by observing which intermediates were consumed and which ones were accumulated following the addition of inhibitors. The importance of Krebs's work was acknowledged when he was awarded the Nobel Prize in 1953, together with Fritz Lipmann who discovered a coenzyme that is an important electron carrier molecule in the electron transport chain.

In the 1920s and 1930s, many scientists worked independently to determine the steps involved in electron transport. In 1961, British biochemist Peter Mitchell proposed a mechanism called chemiosmosis as the process during oxidative phosphorylation that requires a hydrogen ion gradient for the production of ATP. Mitchell carried out most of his research using bacteria. His contribution to cellular respiration was recognized with a Nobel Prize in 1978.

You will use MitochondriaLab to simulate some of the pioneering experiments—similar to those carried out by Szent-Gyorgi, Krebs, Mitchell, and other researchers—that were central to our understanding of cellular respiration. You will add substrates and inhibitors to a reaction flask containing isolated preparations of mitochondria and measure oxygen consumption as a function of the rate of these reactions. The challenge of MitochondriaLab is to carefully analyze your results to develop an understanding of the sequence and control of these reactions.

References
1. Campbell, M. K. *Biochemistry*, 2nd ed. NY: Saunders, 1995.

2. Mathews, C. K., and van Holde, K. E. *Biochemistry*, 2nd ed. Menlo Park, CA: Benjamin/Cummings, 1996.

Introduction

In this laboratory, you will perform simulations of biochemistry experiments designed to study the reactions of aerobic cellular respiration as it occurs in vitro using isolated preparations of mitochondria. By changing reaction conditions through adding substrates and metabolic inhibitors, you will simulate some of the experiments that were used to determine the metabolic reactions and the sequence of the reactions that occur during the Krebs cycle and the electron transport chain.

Objectives

The purpose of this laboratory is to:

- Demonstrate how studying the effects of intermediates in a reaction can be used to determine the steps in a metabolic pathway and the sequence of each enzymatic step in the pathway.
- Demonstrate how changes in substrates, products, and other reaction parameters can affect the reactions involved in the Krebs cycle, electron transport chain, and oxidative phosphorylation.
- Simulate the effects of substances that inhibit the reactions of aerobic cellular respiration.

Before You Begin: Prerequisites

Before beginning MitochondriaLab you should be familiar with the following concepts:

- The importance and functions of enzymes as biological catalysts, basic principles of metabolic pathways, and mechanisms involved in regulating the catalytic activity of an enzyme (see Campbell, N. A., Reece, J. B., and Mitchell, L. G. *Biology* 5/e, and Campbell, N. A., and Reece J. B., *Biology* 6/e, chapter 6).
- The structure and function of the mitochondrion (chapters 7 and 9).
- The reactions of aerobic cellular respiration including glycolysis, the Krebs cycle, the electron transport chain, and oxidative phosphorylation. Be able to describe the primary substrates required, reactions involved, and products generated by each of these reactions (chapter 9).
- The enzymatic steps involved in the Krebs cycle, and the organic acids produced as intermediates during the Krebs cycle (chapter 9).
- Feedback mechanisms involved in the control of cellular respiration (chapter 9).

Assignments

The first screen that appears in MitochondriaLab shows you a biochemistry lab containing all the equipment you will need to perform your experiments. Click on each piece of equipment to learn more about its purpose. For your ease in completing each assignment, the background text relevant to the experiment that you will perform is *italicized*, instructions for each assignment are indicated by plain text, and questions or activities that you will be asked to provide answers for are indicated by **bold text**.

The following assignment is designed to help you become familiar with the operation of MitochondriaLab.

Assignment 1: Getting to Know MitochondriaLab - Setting Up an Experiment

When you enter MitochondriaLab, you are provided with a reaction flask in which you will perform your reactions. These reaction flasks are a standard piece of glassware in a biochemistry lab because they allow biochemists to carefully control and measure reaction conditions within the flask. To begin each experiment, you will start with an empty reaction flask into which you will add isolated preparations of mitochondria. In real-life applications, these are typically mitochondria that have been isolated from mitochondria-rich animal tissues, such as liver or muscle, using cell fractionation techniques. Notice that you have two racks of test tubes to choose from. One rack of tubes contains seven different substrates that you can add to your flask (ADP, ascorbate/TMPD [tetramethyl-p-phenylene diamine], fumarate, glutamate, malate, pyruvate, and succinate). The second rack of tubes contains six different metabolic inhibitors of cellular respiration (antimycin, cyanide, DNP [2,4-dinitrophenol], malonate, oligomycin, and rotenone).

Notice that the chart recorder is connected to an oxygen-measuring electrode that inserts into a port on the left side of the reaction chamber. Substrates and inhibitors will be added to the flask through the port on the right side of the flask. Oxygen consumption in the flask will be determined as a measurement of the progress of reactions in the flask. The chart recorder will measure and plot oxygen concentration in the flask during the time course of your experiments.

1. **Begin your first experiment by developing a hypothesis to predict what will happen to oxygen consumption in the reaction flask after the addition of pyruvate. Develop a second hypothesis to predict how oxygen consumption will change in the flask upon the addition of pyruvate and ADP.** Recall that pyruvate is an end-product of the reactions of glycolysis. Once you have formulated your hypotheses, setup your simulated experiment as follows:

   a. Click on the To Experiment button in the lower left corner of the screen to enter MitochondriaLab. To begin an experiment you, must first add mitochondria to the reaction flask. Do this by clicking on the Add Mitochondria button located below the reaction flask. A running ledger of the components that you added and the time that each component was added to the flask is shown in the lower right corner of the screen. Notice that the mitochondria have been added to a 2.0 ml suspension that consists of a solution buffered to the proper pH for these reactions. This solution contains ions and other components that are required by mitochondria to undergo respiration.

   The chart recorder at the left of the screen will be used to monitor the rate of cellular respiration by the mitochondria in the flask by plotting oxygen concentration (as a percentage of the starting concentration) on the x-axis against time in minutes on the y-axis. Oxygen concentration in the flask begins at 100%. Notice that the chart recorder is plotting oxygen consumption in blue ink. If desired, you can change the recording speed of the simulation by using the small buttons at the upper left corner of the screen. The default value is 4×.

b. **Allow the experiment to proceed for two minutes. Take note of any changes in oxygen consumption that occur during this time. Is this what you expected? Explain your observations.** After two minutes, add pyruvate to the reaction flask by clicking on pyruvate in the substrates box (pyruvate should now be highlighted) and then clicking the Add button. Notice that you added 20 µl of 500 mM pyruvate to the flask. **Follow oxygen consumption in the flask for three minutes. What did you observe during this time? Based on what you already know about the reactions of cellular respiration, explain your observations. Was this result what you predicted based on your hypothesis? Why or why not?**

c. **Add ADP to the flask and follow oxygen consumption. What happened? Was this result what you expected based on your hypothesis? Why or why not. Explain your observations. Explain why oxygen consumption changes so dramatically following the addition of ADP while the addition of pyruvate alone results in slower consumption of oxygen.**

d. Click on the View Chart button to see an enlarged view of the plot. Use the slider bar at the right of the plot to move up or down the plot. Notice that the chart recorder labels the time when you added each component, the concentration of pyruvate added, and the concentration of oxygen (measured in nanomoles of oxygen molecules) at the time the pyruvate was added. You can export the values from this plot by clicking on the Add to Notes button in the lower left corner of the screen. A new browser window will appear with the data in your notebook. You can now print your data from this window or save your data to a disk.

e. Click on the Return to Experiment button to return to the experiment in progress. Reset the reaction flask by clicking on the Clean and Reset Chamber button below the flask to prepare the reaction flask for another experiment. **Formulate a hypothesis to explain what you think will happen when you add succinate to the flask. Run the experiment by adding mitochondria to the clean flask, allowing the reaction to proceed for one minute, and then adding succinate to the flask. Did the results of this experiment validate your hypothesis? Why or why not? Explain why the addition of succinate produced the observed effect on oxygen consumption.**

f. Repeat the procedure in step (e) to perform individual experiments for each of the following substrates: fumarate, ascorbate/TMPD (tetramethyl-*p*-phenylene diamine), malate, and ADP. *Note*: Ascorbate and TMPD are synthetic electron donor molecules that can supply electrons to the electron transport chain. **For each experiment, save all data to your notebook so** you can compare results. Did the effect on oxygen consumption appear to be the same for each substrate? Describe any obvious differences in oxygen consumption that you observed with the different substrates. Based on what you know about the purpose of each substrate in the reactions of cellular respiration, provide possible explanations for the differences in oxygen consumption that you observed. What differences in oxygen consumption did you observe with the use of ADP as a substrate alone compared with using pyruvate and ADP (step c)? Explain these differences.

Assignment 2: The Catalytic Effect of Intermediates and Inhibitors on the Krebs Cycle

One of the important concepts established by many of Krebs's experiments was that the organic acids produced during the citric acid cycle can stimulate oxygen consumption. In addition, Krebs noted that these intermediates are generated in cyclical fashion and that each acid serves as the substrate for the next enzymatic reaction in the cycle. Therefore, once one organic acid is formed, the rate of subsequent reactions is dependent on the production of this organic acid. By following the production and catabolism of organic acids in the pathway it was possible to determine the sequence of enzymatic steps involved in the Krebs cycle.

The use of metabolic inhibitors was an essential aspect of the experiments which determined that the reactions of the Krebs cycle are cyclical. By adding an inhibitor and then measuring the accumulation of an intermediate, it is possible to determine the order or sequence of reactions in the cycle. The following assignment is designed to help you determine the sequence of reactions involved in the Krebs cycle by studying the effects of inhibitors on oxygen consumption.

1. *Malonate is an analog of one of the organic acids that is naturally produced during the Krebs cycle. Malonate acts as an inhibitor of the Krebs cycle. Your job is to determine which enzyme in the cycle is inhibited by malonate.* Begin an experiment by adding mitochondria and malonate. Allow the reaction to proceed for one minute then add succinate. **What happens to oxygen consumption in the experiment?** To determine which step in the Krebs cycle is inhibited by malonate, try adding pyruvate to the experiment. Perform similar independent experiments by adding malate, fumarate, and ADP, to mitochondria containing malonate. Try other substrates as well. **For each experiment describe what happens to oxygen consumption? What do these result tell you about the sequence of succinate, pyruvate, malate and fumarate relative to where malonate is acting?** Based on your results, propose a order of reactions in which each substrate appears and suggest a step in the Krebs cycle that is inhibited by malonate. Consult a textbook if needed to verify your answer and to identify the enzyme inhibited by malonate.

    a. **Once you have determined malonate's site of action, explain what would happen to the concentration of each of the following molecules: citrate, isocitrate, α-ketoglutarate, succinate, oxaloacetate, malonate,**

ADP, and ATP in the experiment if you treated these mitochondria with excess amounts of malonate?

Assignment 3: The Effect of Inhibitors on Oxidative Phosphorylation

DNP (2,4-dinitrophenol) and oligomycin are examples of inhibitors called uncoupling agents. Uncouplers inhibit ATP synthesis but allow the other reactions of cell respiration to proceed normally. **Develop a hypothesis to predict the effects of each agent on oxygen consumption then test the effects of each uncoupler as follows:**

1. Begin an experiment with mitochondria, pyruvate, and ADP. Allow the reaction to proceed for one minute, then add an excess amount of DNP by clicking on DNP and then clicking twice on the Add button. **What happens to oxygen consumption? Will the addition of more ADP influence oxygen consumption? Add ADP and observe what happens. Explain your answers. How might DNP be acting on the reactions of oxidative phosphorylation to cause the change in oxygen consumption that you observed?**

    a. **What do you think is happening to the concentration of H^+, ADP, and ATP in this experiment? Repeat this experiment using oligomycin. What did you observe? Refer to a biochemistry textbook or discuss your results with your instructor to learn how each uncoupling agent disrupts oxidative phosphorylation.**

2. To better understand the coupling effect, and the actions of DNP and oligomycin, perform an experiment with mitochondria, pyruvate, and ADP. Allow this experiment to proceed for two minutes, then add oligomycin. **What happened to oxygen consumption during this time? Follow this experiment for two or three minutes. Add ADP to the experiment. What happens? Try adding DNP. What happens to oxygen consumption now? Explain each of these results. Why did ADP and DNP each produce the effects that you saw? Compare results from this and the previous experiment. What can you propose about the possible actions of each of these inhibitors on the reactions of oxidative phosphorylation? Consult a textbook to identify the site of action for each uncoupling agent.**

Assignment 4: Studying the Efficiency of Oxidative Phosphorylation by Measuring the Rate of Oxygen Consumption

Biochemists who study the efficiency of oxidative phosphorylation in isolated preparations of mitochondria are interested in examining the amount of ATP produced relative to the total amount of energy generated during the oxidation of substrates in reactions of cellular respiration. Recall, however, that ATP production depends on the electron flow produced by oxidizing substrates as well as the concentration of ADP in the mitochondrion. Because oxidative phosphorylation relies on this electron flow as well as ADP, oxidative phosphorylation is said to be "coupled" to respiration (electron flow from substrates to oxygen). One way to

26

measure the efficiency of coupling is to determine what is known as a P/O ratio. The P/O ratio is a measure of the number of ATP molecules that are synthesized as a result of the energy from one pair of electrons that completes the chain by being transferred to one atom of oxygen. A P/O ratio can be calculated by dividing the number (in nmoles) of molecules of ATP synthesized (or ADP consumed) by the number of oxygen atoms (in nmoles) consumed. Depending on the substrates, inhibitors, oxygen concentration, and other parameters, biochemists often categorize oxygen consumption in isolated mitochondria into five energy states. Each state represents different rates of oxygen consumption depending on the components added to the preparation of mitochondria.

Generally, a P/O ratio of approximately 3 (3 moles of ATP) is expected from the oxidation of one mole of NADH. This value is a measure of oxygen consumption during ADP phosphorylation in the ATP-synthesizing step of oxidative phosphorylation catalyzed by ATP synthase. The P/O ratio for any given substrate can be measured. A P/O ratio indicates that a substrate has been oxidized to produce NADH, and the value of the ratio depends on where the electrons from a given substrate or from NADH enter (known as coupling) the electron transport chain. When substrates donate electrons that enter early in the chain (further from oxidative phosphorylation), these electrons are involved in more transfer reactions. As a result, these electrons contribute more to the H^+ gradient needed for oxidative phosphorylation than electrons coupled to the chain at later stages (thus resulting in a high P/O ratio).

The following exercise is designed to help you understand the effects of different substrates on ADP and oxygen consumption by measuring ATP-to-oxygen ratios in your experiments. You will be measuring the amount of ADP that was phosphorylated to ATP to determine P/O ratios. To calculate a P/O ratio, you need to know the following parameters: (1) the reaction is performed at 25°C, (2) the volume of the reaction flask is 2 ml, and (3) there are 237 nanomoles (nmoles) of molecular oxygen (O_2)/ml of reaction volume at 25°C.

1. *For this assignment, we will consider two energy states of mitochondria. State 4 represents the rate of oxygen consumption due to the substrate alone, while state 3 represents oxygen consumption during ADP phosphorylation.* Begin an experiment by adding mitochondria, pyruvate, and ADP. Follow this experiment for several minutes.

   a. The mitochondria will go into state 3 when ADP is added. Click on the View Chart button. Note that once all of the ADP has been phosphorylated to ATP this will be labeled on the plot as ADP phosphorylated. At this point, the mitochondria are now in state 3. Recall that the oxygen recorder is plotting molecular oxygen concentration as a percent of the starting concentration (100%). To determine the amount of molecular oxygen consumed while the mitochondria were in state 3, subtract the percent of oxygen present at the time when ADP was fully phosphorylated from the amount of oxygen present when ADP was first added to the experiment. Note: For your convenience, use the calculator in MitochondriaLab. Follow the steps below to determine the P/O ratio for pyruvate during state 3.

(1) To determine the number of nmoles of molecular oxygen present in the reaction flask, multiply 237 nmoles/ml molecular oxygen by 2 ml (the size of the reaction flask). Because P/O values are calculated based on the amount of atomic oxygen consumed, you must now convert nmoles of molecular oxygen into nmoles of atomic oxygen by multiplying nmoles of molecular oxygen in the reaction flask by 2 (recall that there are two oxygen atoms in each molecule of oxygen). This value now represents nmoles of atomic oxygen in the flask.

(2) To determine nmoles of atomic oxygen consumed in state 3, multiply the percent of molecular oxygen consumed in state 3 by the number of nmoles of atomic oxygen in the flask. (For example, if you observed that 5% of oxygen was consumed during state 3, multiply 0.005 by the number of nmoles of atomic oxygen in the flask that you calculated in step (1) above.) This value represents nmoles of atomic oxygen consumed in state 3.

b. To calculate the P/O ratio in state 3, you must also determine the amount of ADP in nmoles. Recall that you added 20 µl of 10 mM ADP to the reaction flask. Convert this to nmoles first. To determine the P/O ratio, divide nmoles of ADP added by nmoles of atomic oxygen consumed (determined in step 2).

c. **Repeat this experiment using malate, succinate, and ascorbate/TMPD as the substrates. Calculate P/O ratios for each substrate in state 3 and compare these ratios. Refer to a textbook to use diagrams of the Krebs cycle and electron transport chain to trace each substrate to where the substrate itself or electrons from its oxidation enter (are coupled to) the reactions of cellular respiration. Do the ratios make sense to you?**

d. As you learned in the first assignment, Getting to Know MitochondriaLab, ascorbate/TMPD is not a natural substrate combination. From the P/O ratio, you should be able to narrow down the possible location at which TMPD donates electrons to the reactions of cellular respiration.

2. Repeat experiment 1 with glutamate and then fumarate. **Are the state 3 ratios different for the two substrates?** You will have to look up the glutamate dehydrogenase reaction to determine whether the P/O ratio makes sense, because glutamate is not a direct substrate of the Krebs cycle; it enters the reactions of cellular respiration at a different stage. **By what pathway does glutamate donate energy to the reactions of cellular respiration? Trace the path of energy from fumarate. What is the difference between the glutamate and fumarate pathways? What is the rate-limiting step in each reaction pathway?**

Assignment 5: Group Assignments

Certain poisons act as metabolic inhibitors by binding to molecules in the electron transport chain, thereby preventing electron transport and blocking production of the H^+ gradient required for oxidative phosphorylation. Binding can occur in a reversible or irreversible fashion. Rotenone is a poison that works as an irreversible inhibitor of the electron transport chain. Rotenone is frequently used to selectively kill undesirable species of insects and nonnative fish, such as grass carp, that can cause extensive destruction of plant species in lakes and reservoirs.

The following exercises are designed to help you determine which complex of the electron transport chain is a possible binding site for rotenone and other metabolic inhibitors, as well as examine where other substrates enter the electron transport chain. Before you begin these assignments, you may need to refer to a biochemistry textbook to view a detailed figure of the electron transport chain complexes I, II, III, and IV, and the electron transport molecules involved in each complex. Work together in a group of four students to complete the following assignments.

1. Begin an experiment with mitochondria, ADP, and glutamate. Allow the experiment to proceed for one minute, then add rotenone. Note: Keep a record of all your experiments in your lab notebook so you can refer back to your results as you interpret the results from other experiments. **What happened to oxygen consumption? Now add ADP to the flask. What happened to oxygen consumption after adding ADP? Is this what you expected? Explain your answer.**

    a. Ascorbate can bind to a complex in the electron transport chain and donate electrons to the chain. Use ascorbate to help you pinpoint where rotenone is blocking the chain by repeating the same experiment with mitochondria, ADP, and glutamate; then wait one minute, add rotenone, wait another minute, and then add ADP and ascorbate/TMPD to the experiment. **What happened to oxygen consumption? Can you determine which complex of the electron transport chain is bound by ascorbate? Based on these results and knowledge of where ascorbate binds, can you determine which complex (I, II, III, or IV) is likely to be inhibited by rotenone?** Continue to the next experiment to pinpoint the site of action of ascorbate and rotenone.

    b. Succinate can also bind to a complex in the electron transport chain and donate electrons to the chain. Use succinate to help you determine where rotenone is blocking the chain by repeating the same experiment as above with mitochondria, ADP, and glutamate; then wait one minute, add rotenone, wait another minute, and then add ADP and succinate to the experiment. **What happened to oxygen consumption? Can you determine which complex of the electron transport chain is bound by succinate? Based on these results and knowledge of where succinate binds, can you determine which complex (I, II, III, or IV) is likely to be inhibited by rotenone.** Refer to a biochemistry textbook if necessary to determine which complex of the electron transport chain is bound by succinate. **Based on your results and knowledge of where succinate binds, which complex (I, II, III, or IV) is likely to be inhibited by rotenone?** Continue to the next experiment to pinpoint the site of action of rotenone, ascorbate, and succinate.

c. Antimycin is another metabolic inhibitor that acts on the electron transport chain. Use antimycin to help you pinpoint where rotenone is blocking the chain by repeating the experiment in step (b) but after adding rotenone, add ADP and antimycin to the experiment. **What happened to oxygen consumption? Which complex of the electron transport chain is bound by antimycin? Based on your results and knowledge of where antimycin binds, which complex (I, II, III, or IV) is likely to be inhibited by rotenone? Are all of your results consistent enough to help you pinpoint the binding site of rotenone? Once you have determined which complex is bound by rotenone, discuss your answer with your instructor to find out the specific molecules affected by rotenone, ascorbate, succinate, and antimycin.**

2. Repeat experiment (b) above, but after adding succinate and ADP, wait one minute and then add cyanide as a final step in the experiment. Cyanide blocks the electron transport chain in the same fashion as carbon monoxide (CO), the colorless, odorless gas produced when many organic materials are burned. **What happened to oxygen consumption in this experiment? Repeat the same experiment for each of the other experiments you conducted above. For each experiment, what happened to oxygen consumption? Is this what you expected? Why or why not? Can you add substrates to bypass the cyanide block? Try this experiment with all of the substrates available. What did you observe? Explain your results, then determine which complex of the electron transport chain is blocked by cyanide.**

3. **Based on what you know about the electron transport chain and oxidative phosphorylation, explain how adjusting the pH of your reaction flask can be used to artificially bypass the effects of these inhibitors. What would you add to the reaction flask if you were actually going to perform this experiment? Based on what you know about the actions of oligomycin from earlier experiments, how might this experiment be affected by the addition of oligomycin?**

FlyLab

Background

While developing his theory of evolution by natural selection, Charles Darwin was unaware of the molecular basis for evolutionary change and inheritance. Around the same time that Darwin was making his landmark observations that led to his publication, *The Origin of Species,* the Augustinian monk Gregor Johann Mendel was performing experiments with garden peas that dramatically influenced the field of biology. Mendel's interpretation of his experimental results served as the foundation for the discipline known as **genetics**. Genetics is the discipline of biology concerned with the study of heredity and variation. In 1866, Mendel published a classic paper in which he described phenomena in garden peas that laid down the principles for genetic inheritance in living organisms.

Our understanding of genetics in higher organisms such as humans has advanced considerably since Mendel's work. This has occurred, in part, because of significant advances in our understanding of the molecular biology of living cells. The molecular biology revolution that led to our current understanding of genetics was greatly accelerated in 1953 by James Watson and Francis Crick, who revealed the structure of DNA as a double-helical molecule. Considering that a detailed understanding of the structure of DNA, chromosomes and genes, and the events of meiosis were not known during Mendel's time, Mendel's work was a particularly incredible accomplishment. Approximately 40 years passed before the significance of Mendel's insight was realized. Mendel's work gained acceptance as his experiments were replicated and publicized by scientists who performed genetic experiments several years after Mendel died. Mendel's landmark principles on inheritance continue to form the basis for our current understanding of genetics in living organisms. As a modern-day student studying biology, having the benefit of hindsight by studying gene and chromosome structure, and understanding how **gametes** form by meiosis, is a great advantage in developing an understanding of inheritance.

The experimental model for Mendel's work involved performing cross-pollination experiments with a strain of garden peas called *Pisum sativum*. Mendel was interested in studying the inheritance of a number of different characters, or heritable features, of *Pisum*. He considered seven different characters including flower color, flower position, seed color, seed shape, pod color, pod shape, and stem length. Variations of a given character are known as traits. For example, when studying flower color as a character, Mendel traced the inheritance of two traits for flower color, purple flowers and white flowers. Many of the basic genetic principles established by Mendel arose from his observations of the results produced by a simple cross-pollination experiment called a **monohybrid cross**. In a monohybrid cross, individuals with one pair of contrasting traits for a given character are mated. For example, a plant with purple flowers is mated with a plant that has white flowers.

Mendel proposed several basic principles to explain inheritance. One of his first principles was that characters are determined by what he described as "heritable factors" or "units of inheritance." The factors that Mendel described are what we now call genes. Mendel noted that alternative forms of a gene, what we now call **alleles**, are responsible for variations in inherited characters. For example, in *Pisum sativum* there are two alleles for flower color, a white flower allele and a purple flower allele. Mendel also proposed that certain alleles, called **dominant alleles**, are always expressed in the appearance of an organism and that the expression or appearance of other alleles, called **recessive alleles**, was sometimes hidden or masked. Recall that modern-day geneticists frequently use capital letters to indicate dominant genes and lowercase letters to indicate recessive genes. The appearance of a particular trait is referred to as the **phenotype** of an organism while the genetic composition of an organism is known as the organism's **genotype**. Mendel established that an organism typically inherits two alleles for a given character, one from each parent. Using modern genetic terminology, we say that an organism that contains a pair of the same alleles is **homozygous** for a particular trait while an organism that contains two different alleles for a trait is **heterozygous** for that trait. Mendel also postulated that when an organism contains a pair of units, the units separate, or segregate, during gamete formation so that each individual gamete produced receives only one unit of the pair. This postulate became known as Mendel's **law of segregation of alleles**.

One example of a monohybrid cross performed by Mendel involved the inheritance of pea flower color. Mendel began by crossing true-breeding homozygous parents called the **P generation**. He crossed a plant with purple flowers with a plant showing white flowers. The first offspring produced from such a cross, called the F_1 **generation**, all showed purple flowers. The results of this cross were explained when Mendel determined that the F_1 plants were heterozygotes and that these plants showed purple flowers because the purple flower allele was dominant and the white flower allele was recessive. Self-pollination of F_1 plants produced a second generation of plants called the F_2 **generation**, in which the phenotypic ratio of purple-flowered plants to white-flowered plants was approximately 3:1. As he performed monohybrid crosses for several other characters, Mendel discovered that this ratio was characteristic of a cross between heterozygotes. The genotypes and phenotypes that may result from a genetic cross can be predicted by using a Punnett square. You should be familiar with constructing Punnett squares to predict the results of monohybrid and **dihybrid crosses**. Mendel used dihybrid crosses to explain his **law of independent assortment**, in which he postulated that alleles for different characters (for example, pea color and pea shape) segregate into gametes independently of each other. A dihybrid cross between F_1 heterozygotes reveals a phenotypic ratio of approximately 9:3:3:1 in the F_2 generation.

The 3:1 ratio predicted for Mendel's monohybrid cross and the 9:3:3:1 phenotypic ratio predicted for a dihybrid cross are hypothetical expected ratios. When performing a real genetic cross based on Mendelian principles, however, such a cross is subject to random changes and experimental errors that produce chance deviation in the actual phenotypic ratios that one may observe. To accurately evaluate genetic inheritance, it is essential that observed deviation in a phenotypic ratio be determined and compared with predicted ratios.

Chi-square analysis (χ^2) is a statistical method that can be used to evaluate how observed ratios for a given cross compare with predicted ratios. Chi-square analysis considers the chance deviation for an observed ratio, and the sample size of the offspring, and expresses these data as a single value. Based on this value, data are converted into a single probability value (p-value), which is an index of the probability that the observed deviation occurred by random chance alone. Biologists generally agree on a p-value of 0.05 as a standard cutoff value, known as the level of significance, for determining if observed ratios differ significantly from expected ratios. A p-value below 0.05 indicates that it is unlikely that an observed ratio is the result of chance alone. When we predict that data for a particular cross will fit a certain ratio—for example, expecting a 3:1 phenotypic ratio for a monohybrid cross between heterozygotes—this assumption is called a **null hypothesis**. Chi-square analysis is important for determining whether a null hypothesis is an accurate prediction of the results of a cross. Based on a p-value generated by chi-square analysis, a null hypothesis may either be rejected or fail to be rejected. If the level of significance is small ($p < 0.05$), it is unlikely (low probability) that the observed deviation from the expected ratio can be attributed to chance events alone. This means that your hypothesis is probably incorrect and that you need to determine a new ratio based on a different hypothesis. If, however, the level of significance is high ($p > 0.05$), then there is a high probability that the observed deviation from the expected ratio is simply due to random error and chi-square analysis would fail to reject your hypothesis.

It is important to realize that the principles established by Mendel can be easily explained by understanding the chromosomal basis of inheritance. For example, in humans, **somatic cells** contain the **diploid number** ($2n$) of chromosomes, which consists of 46 chromosomes organized as 23 pairs of **homologous chromosomes** (homologues). Chromosomes 1 through 22 are called **autosomes** and the twenty-third pair of chromosomes are called **sex chromosomes**. During meiosis, gamete formation leads to the formation of sex cells that contain a single set of 23 chromosomes—the **haploid number** (n) of chromosomes. Because chromosomes are present as pairs in human cells, genes located on each chromosome are also typically present as pairs. Mendel's law of segregation of alleles is explained by the separation of each pair of homologous chromosomes that occurs during meiosis resulting in each individual gamete receiving only one copy of each chromosome. This law applies to, and is explained by, the separation of **unlinked genes** because these genes are located on different chromosomes. As a result, during meiosis the chromosomes align at the metaphase plate and separate in a random, independent fashion.

An expansion in our knowledge of Mendelian genetics has led to a detailed understanding of different types of inheritance events such as codominance, incomplete dominance, linked inheritance, inheritance of multiple alleles, lethal mutations, and the genetic transmission of a number of different genetic disorders. For many reasons, Mendel's pea plants served as an ideal model organism for him to study. In fact, pea plants allowed Mendel to succeed where others failed. Peas are easy to grow, they cross-fertilize and self-fertilize easily, and they mature quickly. More recently, a number of other organisms have served similar roles as model organisms for scientists who study genetics. Some of these include the flowering plant *Arabidopsis thaliana,* a common inhabitant of home aquariums called the striped zebrafish (*Brachydanio rerio*), many species of mice, and the common fruit fly, *Drosophila melanogaster.*

In particular, *Drosophila* has been one of the most well studied model organisms for learning about genetics and embryo development. These small flies are hardy to grow under lab conditions, and they reproduce easily with a relatively short life cycle compared with vertebrate organisms; hence, crosses can be performed and offspring counted over reasonable intervals of time. Another advantage of *Drosophila* is that the **loci** for many genes on the four chromosomes in the fly's genome have been determined, and a very large number of mutations of the **wild-type** fly have been developed that effect different phenotypes in *Drosophila*. Because of some of these characteristics, *Drosophila* has an important historical place in the field of genetics and continues to be an important model organism for studying genetic inheritance that universally applies to most organisms. Using FlyLab, you will design and carry out experimental crosses using *Drosophila melanogaster*. In the future, you will have the opportunity to use PedigreeLab to study inheritance of different genetic disorders through several generations of humans.

References

1. Klug, W. S., and Cummings, M. R. *Essentials of Genetics*, 2nd ed. Upper Saddle River, NJ: Prentice Hall, 1996.

2. Lindsley, D. and Zimm, G. *The Genome of Drosophila melanogaster.* San Diego: Academic Press, 1992.

3. Orel, V. *Mendel.* New York: Oxford University Press, 1984.

4. Soudek, D. "Gregor Mendel and the People Around Him." *Am. J. Hum. Genet.,* May 1984.

5. Stern, C., and Sherwood, E. R. *The Origin of Genetics: A Mendel Source Book.* San Francisco: W. H. Freeman, 1966.

Introduction

FlyLab will allow you to play the role of a research geneticist. You will use FlyLab to study important introductory principles of genetics by developing hypotheses and designing and conducting matings between fruit flies with different mutations that you have selected. Once you have examined the results of a simulated cross, you can perform a statistical test of your data by chi-square analysis and apply these statistics to accept or reject your hypothesis for the predicted phenotypic ratio of offspring for each cross. With FlyLab, it is possible to study multiple generations of offspring, and perform testcrosses and backcrosses. FlyLab is a very versatile program; it can be used to learn elementary genetic principles such as dominance, recessiveness, and Mendelian ratios, or more complex concepts such as sex-linkage, epistasis, recombination, and genetic mapping.

Objectives

The purpose of this laboratory is to:

- Simulate basic principles of genetic inheritance based on Mendelian genetics by designing and performing crosses between fruit flies.
- Help you understand the relationship between an organism's genotype and its phenotype.
- Demonstrate the importance of statistical analysis to accept or reject a hypothesis.
- Use genetic crosses and recombination data to identify the location of genes on a chromosome by genetic mapping.

Before You Begin: Prerequisites

Before beginning FlyLab you should be familiar with the following concepts:

- Chromosome structure, and the stages of gamete formation by meiosis (see Campbell, N. A., Reece, J. B., and Mitchell, L. G. *Biology* 5/e, and Campbell, N. A., and Reece J. B., *Biology* 6/e, chapters 13–15).
- The basic terminology and principles of Mendelian genetics, including complete and incomplete dominance, epistasis, lethal mutations, recombination, autosomal recessive inheritance, autosomal dominant inheritance, and sex-linked inheritance (chapters 14 and 15).
- Predicting the results of monohybrid and dihybrid crosses by constructing a Punnett square (chapter 14)
- How genetic mutations produce changes in phenotype, and beneficial and detrimental results of mutations in a population (chapter 17).

Assignments

To begin an experiment, you must first design the phenotypes for the flies that will be mated. In addition to wild-type flies, 29 different mutations of the common fruit fly, *Drosophila melanogaster*, are included in FlyLab. The 29 mutations are actual known mutations in *Drosophila*. These mutations create phenotypic changes in bristle shape, body color, antennae shape, eye color, eye shape, wing size, wing shape, wing vein structure, and wing angle. For the purposes of the simulation, genetic inheritance in FlyLab follows Mendelian principles of complete dominance. Examples of incomplete dominance are not demonstrated with this simulation. A table of the mutant phenotypes available in FlyLab can be viewed by clicking on the Genetic Abbreviations tab which appears at the top of the FlyLab homepage. When you select a particular phenotype, you are not provided with any information about the dominance or recessiveness of each mutation. FlyLab will select a fly that is homozygous for the particular mutation that you choose, unless a mutation is lethal in the homozygous condition in which case the fly chosen will be heterozygous. Two of your challenges will be to determine the zygosity of each fly in your cross and to determine the effects of each allele by analyzing the offspring from your crosses.

One advantage of FlyLab is that you will have the opportunity to study inheritance in large numbers of offspring. FlyLab will also introduce random experimental deviation to the data as would occur in an actual experiment! As a result, the statistical analysis that you will apply to your data when performing chi-square analysis will provide you with a very accurate and realistic analysis of your data to confirm or refute your hypotheses.

For your ease in completing each assignment, the background text relevant to the experiment that you will perform is *italicized*, instructions for each assignment are indicated by plain text, and questions or activities that you will be asked to provide answers for are indicated by **bold text**.

The following assignment is designed to help you become familiar with the operation of FlyLab.

Assignment 1: Getting to Know FlyLab: Performing Monohybrid, Dihybrid, and Trihybrid Crosses

1. To begin a cross, you must first select the phenotypes of the flies that you want to mate. Follow the directions below to create a monohybrid cross between a wild-type female fly and a male fly with sepia eyes.

    a. To design a wild-type female fly, click on the Design button below the gray image of the female fly. Click on the button for the Body Color trait (or any trait) on the left side of the Design view. The small button next to the words "Wild Type" should already be selected (bolded). To choose this phenotype, click the Select button below the image of the fly at the bottom of the design screen. Remember that this fly represents a true-breeding parent that is homozygous for wild type alleles. The selected female fly now appears on the screen with a "+" symbol indicating the wild-type phenotype.

    b. To design a male fly with sepia eyes, click on the Design button below the gray image of the male fly. Click on the button for the Eye Color trait on the left side of the Design view. Click on the small button next to the word "Sepia." Note how eye color in this fly compares with the wild-type eye color. Choose this fly by clicking on the Select button below the image of the fly at the bottom of the Design screen. The male fly now appears on the screen with the abbreviation "SE" indicating the sepia eye mutation. This fly is homozygous for the sepia eye allele. These two flies represent the parental generation (P generation) for your cross.

    c. **Based on what you know about the principles of Mendelian genetics, predict the phenotypic ratio that you would expect to see for the F_1 offspring of this cross and describe the phenotype of each fly.**

    d. To select the number of offspring to create by this mating, click on the popup menu on the left side of the screen and select 10,000 flies. To mate the two flies, click on the Mate button between the two flies. Note the fly images that appear in the box at the bottom of the screen. Scroll up to see the parent flies and down to see the wild type offspring. These offspring are the F_1 generation. **Are the phenotypes of the F_1 offspring what you**

would have predicted for this cross? Why or why not? <u>Note</u>: The actual number of F_1 offspring created by FlyLab does not exactly equal the 10,000 offspring that you selected. This difference represents the experimental error introduced by FlyLab.

e. To save the results of this cross to your lab notes, click on the Results Summary button on the lower left side of the screen. A panel will appear with a summary of the results for this cross. Note the number of offspring, proportion of each phenotype and observed ratios for each observed phenotype. Click the Add to Lab Notes button at the bottom of the panel. Click the OK button to close this panel. To comment on these results in your lab notes, click on the Lab Notes button and move the cursor to the space above the dashed line and type a comment such as, "These are the results of the F1 generation for my first monohybrid cross." Click the Close button to close this panel and return to the Mate screen.

f. To set up a cross between two F_1 offspring to produce an F_2 generation, be sure that you are looking at the two wild-type offspring flies in the box at the bottom of the screen. If not, scroll to the bottom of this box until the word "Offspring" appears in the center of the box. Click the Select button below the female wild-type fly image, then click the Select button below the male wild-type fly image. Note that the two F_1 offspring that you just selected appear at the top of the screen as the flies chosen for your new mating. Click on the Mate button between the two flies. The F_2 generation of flies now appears in the box at the bottom of the screen. Use the scroll buttons to view the phenotypes of the F_2 offspring.

g. **Examine the phenotypes of the offspring produced and save the results to your lab notes by clicking on the Results Summary button on the lower left side of the Mate view. Note observed phenotypic ratios of the F_2 offspring.** Click the Add to Lab Notes button at the bottom of the panel. Click the OK button to close the panel.

h. To validate or reject a hypothesis, perform a Chi-square analysis as follows. Click on the Chi-Square Test button on the lower left side of the screen. To ignore the effects of sex on this cross, click on the Ignore Sex button. Enter a predicted ratio for a hypothesis that you want to test. For example, if you want to test a 4:1 ratio, enter a 4 in the first box under the Hypothesis column and enter a 1 in the second box. To evaluate the effects of sex on this cross, simply type a 4 in each of the first two boxes, and type a 1 in each of the last two boxes. Click the Test Hypothesis button at the bottom of the panel. A new panel will appear with the results of the chi-square analysis. **Note the level of significance displayed with a recommendation to either reject or not reject your hypothesis. What was the recommendation from the chi-square test? Was your ratio accepted or rejected?** Click the Add to Lab Notes button to add the results of this test to your lab notes. Click OK to close this panel.

i. To examine and edit your lab notes, click on the Lab Notes button in the lower left corner of the screen. Click the cursor below the recommendation line and type the following: "These are my results for the F_2 generation of my first monohybrid cross. These data do not seem to follow a 4:1 ratio." To print your lab notes, you can export this data table as an html file by clicking on the Export button. In a few seconds, a new browser window should appear with a copy of your lab notes. You can now save this file to disk and/or print a copy of your lab notes. Click the Close button at the bottom of the panel to close the panel.

j. **Repeat the Chi-square analysis with a new ratio until you discover a ratio that will not be rejected. What did you discover to be the correct phenotypic ratio for this experiment? Was this what you expected? Why or why not? What do the results of this experiment tell you about the dominance or recessiveness of the sepia allele for eye color?**

2. Click on the New Mate button in the lower left corner of the screen to clear your previous cross. Following the procedure described above, perform monohybrid crosses for at least three other characters. For each cross, develop a hypothesis to predict the results of the phenotypes in the F_1 and F_2 generations and perform chi-square analysis to compare your observed ratios with your predicted ratios. For each individual cross, try varying the number of offspring produced.

   What effect, if any, does this have on the results produced and your ability to perform chi-square analysis on these data? If any of your crosses do not follow an expected pattern of inheritance, provide possible reasons to account for your results.

3. Once you are comfortable with using FlyLab to perform a monohybrid cross, design a dihybrid cross by selecting and crossing an ebony body female fly with a male fly that has the vestigial mutation for wing size.

   Develop a hypothesis to predict the results of this cross and describe each phenotype that you would expect to see in both the F_1 and F_2 generations of this cross.

   Analyze the results of each cross by Chi-square analysis and save your data to your lab notes as previously described in the assignments for a monohybrid cross.

   Describe the phenotypes that you observed in both the F_1 and F_2 generations of this cross. How does the observed phenotypic ratio for the F_2 generation compare with your predicted phenotypic ratio? Explain your answer.

4. Use FlyLab to perform a trihybrid cross by designing and crossing a wild-type female fly and a male fly with dumpy wing shape, ebony body color, and shaven bristles.

Develop a ~~hypothesis~~ to predict the ~~results~~ of this cross and describe each phenotype that you ~~would expect~~ to see in the F_2 generation of this cross. Perform ~~your cross~~ and evaluate your ~~hypothesis~~ by Chi-square analysis. ~~What~~ was the trihybrid phenotypic ratio produced ~~for the~~ F_2 generation?

Assignment 2: Testcross

*A **testcross** is a valuable way to use a genetic cross to determine the genotype of an organism that shows a dominant phenotype but unknown genotype. For instance, using Mendel's peas, a pea plant with purple flowers as the dominant phenotype could have either a homozygous or a heterozygous genotype. With a testcross, the organism with an unknown genotype for a dominant phenotype is crossed with an organism that is homozygous recessive for the same trait. In the animal- and plant-breeding industries, testcrosses are one way in which the unknown genotype of an organism with a dominant trait can be determined. Perform the following experiment to help you understand how a testcross can be used to determine the genotype of an organism.*

1. Design a female fly with brown eye (BW) color (keep all other traits as wild-type), and design a male fly with ebony body color (E; keep all other traits as wild-type). Mate the two flies. Examine the F_1 offspring from this cross and save your data to your lab notes. Add to your data any comments that you would like.

   To determine the genotype of an F_1 wild-type female fly, design a male fly with brown eye color and ebony body color, then cross this fly with an F_1 wild-type female fly. Examine the results of this cross and save the results to your lab notes.

   What was the phenotypic ratio for the offspring resulting from this testcross? Based on this phenotypic ratio, determine whether the F_1 wild-type female male was double homozygous or double heterozygous for the eye color and body color alleles. Explain your answer. If your answer was double homozygous, describe an expected phenotypic ratio for the offspring produced from a testcross with a double heterozygous fly. If your answer was double heterozygous, describe an expected phenotypic ratio for the offspring produced from a testcross with a homozygous fly.

Assignment 3: Lethal Mutations

Five of the mutations in FlyLab are lethal when homozygous. When you select a lethal mutation from the Design view, the fly is made heterozygous for the mutant allele. If you select two lethal mutations that are on the same chromosome (same linkage group, or the "cis" arrangement), then the mutant alleles will be placed on different homologous chromosomes (the "trans" arrangement). Crosses involving lethal mutations will not show a deficit in the number of offspring. FlyLab removes the lethal genotypes from among the offspring and "rescales" the probabilities among the surviving genotypes. Hence, the total number of offspring will be the same as for crosses involving only nonlethal mutations. Perform the following crosses to demonstrate how Mendelian ratios can be modified by lethal mutations.

1. Design a cross between two flies with aristapedia mutations for antennae shape. Mate these flies.

   What phenotypic ratio did you observe in the F_1 generation? What were the phenotypes? Perform an F_1 cross between two flies with the aristapedia phenotype. What phenotypic ratio did you observe in the F_2 generation? How do these ratios and phenotypes explain that the aristapedia mutation functions as a lethal mutation?

   To convince yourself that the aristapedia allele is lethal in a homozygote compared with a heterozygote, perform a cross between a wild-type fly and a fly with the aristapedia mutation.

   What results did you obtain with this cross?

2. Design a cross between two flies with curly wing shape and stubble bristles.

   Develop a hypothesis to predict the phenotypic ratio for the F_1 generation. Mate these flies. What phenotypic ratio did you observe in the F_1 generation?

   Test your hypothesis by Chi-square analysis. Repeat this procedure for an F_1 cross between two flies that express the curly wing and stubble bristle phenotypes.

   Are the phenotypic ratios that you observed in the F_2 generation consistent with what you would expect for a lethal mutation? Why or why not? Explain your answers.

Assignment 4: Epistasis

*The genetic phenomenon called **epistasis** occurs when the expression of one gene depends on or modifies the expression of another gene. In some cases of epistasis, one gene may completely mask or alter the expression of another gene. Perform the following crosses to study examples of epistasis in* <u>Drosophila</u>.

1. Design and perform a cross between a female fly with vestigial wing size and a male fly with an incomplete wing vein mutation. Carefully study the phenotype of this male fly to be sure that you understand the effect of the incomplete allele.

   What did you observe in the F_1 generation? <u>Note</u>: It may be helpful to click up and down in this display box to closely compare the phenotypes of the F_1 and P generations.

   Was this what you expected? Why or why not? Once you have produced an F_1 generation, mate F_1 flies to generate an F_2 generation.

   Study the results of your F_2 generation, then answer the following questions.

Which mutation is epistatic? Is the vestigial mutation dominant or recessive? Determine the phenotypic ratio that appeared in the dihybrid F_2 generation, and use chi-square analysis to accept or reject this ratio.

2. Perform another experiment by mating a female fly with the apterous wing size mutation with a male fly with the radius incomplete vein structure mutation. Follow this cross to the F_2 generation.

   Which mutation is epistatic? Is the apterous wing mutation dominant or recessive?

Assignment 5: Sex Linkage

For many of the mutations that can be studied using FlyLab, it does not matter which parent carries a mutated allele because these mutations are located on autosomes. Reciprocal crosses produce identical results. When alleles are located on sex chromosomes, however, differences in the sex of the fly carrying a particular allele produce very different results in the phenotypic ratios of the offspring. Sex determination in Drosophila follows an X-Y chromosomal system that is similar to sex determination in humans. Female flies are XX and males are XY. Design and perform the following crosses to examine the inheritance of sex linked alleles in Drosophila.

1. Cross a female fly with a tan body with a wild-type male.

   What phenotypes and ratios did you observe in the F_1 generation?

   Mate two F_1 flies and observe the results of the F_2 generation.

   Based on what you know about Mendelian genetics, did the F_2 generation demonstrate the phenotypic ratio that you expected? If not, what phenotypic ratio was obtained with this cross?

2. Perform a second experiment by crossing a female fly with the vestigial wing size mutation and a white-eyed male. Describe the phenotypes obtained in the F_2 generation. Examine the phenotypes and sexes of each fly.

   Is there a sex and phenotype combination that is absent or underrepresented? If so, which one? What does this result tell you about the sex chromosome location of the white eye allele?

Assignment 6: Recombination

*Mendel's law of independent assortment applies to unlinked alleles, but **linked genes**—genes on the same chromosome—do not assort independently. Yet linked genes are not always inherited together because of **crossing over**. Crossing over, or **homologous recombination**, occurs during prophase of meiosis I when segments of DNA are exchanged between homologous chromosomes.*

*Homologous recombination can produce new and different combinations of alleles in offspring. Offspring with different combinations of phenotypes compared with their parents are called **recombinants**. The frequency of appearance of recombinants in offspring is known as recombination frequency. Recombination frequency represents the frequency of a crossing-over event between the loci for linked alleles. If two alleles for two different traits are located at different positions on the same chromosome (heterozygous loci) and these alleles are far apart on the chromosome, then the probability of a chance exchange, or recombination, of DNA between the two loci is high. Conversely, loci that are closely spaced typically demonstrate a low probability of recombination. Recombination frequencies can be used to develop gene maps, where the relative positions of loci along a chromosome can be established by studying the number of recombinant offspring. For example, if a dihybrid cross for two linked genes yields 15% recombinant offspring, this means that 15% of the offspring were produced by crossing over between the loci for these two genes. A genetic map is displayed as the linear arrangement of genes on a chromosome. Loci are arranged on a map according to map units called centimorgans. One centimorgan is equal to a 1% recombination frequency. In this case, the two loci are separated by approximately 15 centimorgans.*

In Drosophila, unlike most organisms, it is important to realize that crossing over occurs during gamete formation in female flies only. Because crossing over does not occur in male flies, recombination frequencies will differ when comparing female flies with male flies. Perform the following experiments to help you understand how recombination frequencies can be used to develop genetic maps. In the future, you will have the opportunity to study genetic mapping of chromosomes in more detail using PedigreeLab.

1. To understand how recombination frequencies can be used to determine an approximate map distance between closely linked genes, cross a female fly with the eyeless mutation for eye shape with a male fly with shaven bristles. Both of these genes are located on chromosome IV in *Drosophila*. Testcross one of the F_1 females to a male with both the eyeless and shaven bristle traits. The testcross progeny with both mutations or neither mutation (wild-type) are produced by crossing over in the double heterozygous F_1 female. The percentage of these recombinant phenotypes is an estimate of the map distance between these two genes.

   Draw a map that shows the map distance (in map units or centimorgans) between the locus for the shaven bristle allele and the locus for the eyeless allele.

2. To understand how recombination frequencies can be used to determine a genetic map for three alleles, mate a female fly with a black body, purple eyes, and vestigial wing size to a wild-type male. These three alleles are located on chromosome III in Drosophila. Testcross one of the F_1 females to a male with all three mutations. The flies with the least frequent phenotypes should show the same phenotypes; these complementary flies represent double crossovers.

What is the phenotype of these flies? What does this tell you about the position of the purple eye allele compared with the black body and vestigial wing alleles? Sketch a genetic map indicating the relative loci for each of these three alleles, and indicate the approximate map distance between each locus.

Assignment 7: Group Assignment

Work in pairs to complete the following assignment. Each pair of students should randomly design at least two separate dihybrid crosses of flies with mutations for two different characters (ideally choose mutations that you have not looked at in previous assignments) and perform matings of these flies. Before designing your flies, refer to the Genetic Abbreviations chart in FlyLab for a description of each mutated phenotype. Or view the different mutations available by selecting a fly, clicking on each of the different phenotypes, and viewing each mutated phenotype until you select one that you would like to follow. Once you have mated these flies, follow offspring to the F_2 generation.

1. For each dihybrid cross, answer the following questions. Perform additional experiments if necessary to answer these questions.
   a. Which of these traits are dominant and which traits are recessive?
   b. Are any of these mutations lethal in a homozygous fly? Which ones?
   c. Are any of the alleles that you followed sex-linked? How do you know this?
   d. Which alleles appear to be inherited on autosomes?
   e. If any of the genes were linked, what is the map distance between these genes?

2. For at least one of your crosses, attempt to perform the cross on paper using a Punnett square to confirm the results obtained by FlyLab.

3. Ask another pair of students to carry out one of the crosses that you designed. Did they get the same results that you did in the F_1 and F_2 generations? Did they develop the same hypotheses to explain the results of this mating as you did? Explain your answer.

4. Once you have completed this exercise, discuss your results with your instructor to determine if your observations and predictions were accurate.

PedigreeLab

Background

Over a century ago, Gregor Mendel established many of the basic laws of inheritance. In the years following Mendel's discoveries, his work was confirmed and extended by many geneticists. During this time an increased knowledge about DNA structure, and the events of mitosis and meiosis, led to a greater understanding of the chromosomal basis for inheritance which explained many of Mendel's hypotheses. For example, Mendel demonstrated that genes on different chromosomes, or unlinked genes, are inherited by independent assortment. However, when following the inheritance of **linked genes**—genes on the same chromosome—gametes are often produced during meiosis that contain chromosomes with different combinations of alleles compared to the parent chromosomes.

One of the key events of meiosis that has a profound effect on genetic inheritance occurs during prophase of meiosis I. In this stage, homologous chromosomes pair together, in a process called **synapsis**, to form a tetrad. Once paired, a reciprocal exchange or swapping of DNA occurs between segments of DNA on the nonsister chromatids of each homologue. This exchange of DNA is called **crossing over** or **homologous recombination**. Crossing over results in the mixing of alleles between the two homologues. After a single crossover event, each chromosome contains DNA obtained from its homologue. Crossing over can produce new combinations of linked alleles that were not present on the parental chromosomes thus crossing over is one of the important cellular events responsible for the tremendous genetic variation in offspring that arise from gamete formation by meiosis and subsequent sexual reproduction. Offspring that arise from recombinant gametes produced by crossing over are called **recombinants** because they express phenotypes that are different from the parents. Geneticists refer to recombination frequency as a measure of the number of crossover events that occur between two **loci** on a chromosome. On average, one to four crossover events occur between most homologous chromosomes during meiosis in humans. Unfortunately, crossover events can occasionally occur between nonhomologous chromosomes. These events are called **translocations**. Translocations are the underlying molecular cause of certain types of human genetic disorders. For example, one type of leukemia called chronic myelogenous leukemia arises from a translocation of DNA between chromosome 21 and chromosome 14.

The position of DNA exchange between nonsister chromatids during crossing over is called the **chiasma** (pl., chiasmata). Chiasmata can be visualized with an electron microscope as X-shaped configurations where DNA from the homologues overlaps and a complex of enzymes carries out the reactions that result in DNA exchange. Because chiasmata formation and crossing over is, in part, a chance event, the probability of a single crossover event between two loci on a single chromosome is directly proportional to the distance between the two loci. For example, alleles that are far apart on a chromosome are more likely to be involved in crossing over, compared to two closely space loci, because the probability that chiasmata will form between the distant loci is high. Genes that are closer together on the same chromosome will tend to be inherited together because the probability of chiasma formation and crossing over between closely spaced loci is relatively low. Hence, the distance between genes on a chromosome can be estimated by studying the results of genetic linkage studies based on recombination frequencies. Recall that recombinants that result from crossing over between two linked genes can be counted by studying the offspring of a cross between an individual who is a heterozygote for both genes with an individual that is (double) homozygote for both genes. Using recombination frequency values from such a cross, the relative location of the two genes on the same chromosome can be used to develop a genetic map of a chromosome or linkage map, which is an approximate measure of the linear arrangement of loci on a chromosome. Geneticists refer to all of the genes that are connected together on the same chromosome as a linkage group.

In a genetic map, the distance between loci is indicated in map units called centimorgans (cM). One centimorgan is equivalent to a 1% recombination frequency. One centimorgan is equal to approximately 1 megabase (Mb) or one million base pairs of DNA. For two closely spaced loci, maps units are equivalent to the percent of recombinants produced; however, for loci that are far apart, double exchanges of DNA between the two loci (double-crossovers) maintain the allele combinations of the parent effectively reducing the number of recombinants. In this case, map distance is underestimated by comparison of recombination frequencies alone. Map distance must then be measured by multiplying the total number of recombinants by 100 then dividing this value by the total number of offspring.

One of the first geneticists to study recombination, and suggest the use of recombination data as a way to map genes and chromosomes, was Sir Thomas Morgan. In the early 1900s, Morgan was studying inheritance in *Drosophila melanogaster* when he noticed that white-eyed male flies would occasionally appear in a line of flies with wild-type eye color. Morgan hypothesized that the molecular explanation for genetic linkage was the location of a gene on a chromosome, and that the development of recombinant offspring was the result of crossing over during meiosis. A student of Thomas Morgan's, Alfred H. Sturtevant, extended Morgan's observations when he discovered that recombination frequencies between linked genes are additive. For example, when studying recombination frequencies between three linked genes, the location and order of these genes on a chromosome can be arranged in a linear fashion according to the recombination frequency between each locus. The total recombination frequencies between all three loci represents the total length (in centimorgans) occupied by this linkage group.

It is obviously not possible to study the inheritance of linked traits in humans by purposely designing crosses between individuals and carrying out experiments analyzing large numbers of offspring. How then can the inheritance of human genetic disorders be studied? And how can human gene maps and chromosome maps be developed? One technique that is particularly useful for determining the mode of inheritance for a human disorder involves developing a **pedigree**. A pedigree represents a family tree of genetics. In a pedigree, individuals in a family are shown according to their phenotype for a given trait. When a pedigree is developed, symbols are used to signify each family member. A female is represented as a circle and a male is represented as a square. The symbol for an individual who expresses the phenotype for a certain genetic trait is shaded with a particular color. Parents are connected together by a horizontal line while vertical lines from the parents indicate the offspring produced from these parents. Geneticists refer to offspring as *sibs*, an abbreviation for siblings. Sibs are connected to each other by a horizontal line (sibship line). Sibs are shown from left to right in the order in which they are born. In some pedigrees the sibs may be numbered.

By carefully studying the appearance of a particular phenotype in several generations of individuals, especially in a large pedigree, it is often possible to determine the mode of inheritance of that trait. For example, it may be possible to establish whether the trait is inherited by autosomal recessive, autosomal dominant, or sex-linked inheritance. Each of the pedigrees in PedigreeLab traces the inheritance of a single trait. There are limitations to studying inheritance by pedigree analysis. Typically pedigrees tend to follow small numbers of individuals and they may often lack information about certain family members. Although a pedigree is a useful technique for evaluating the mode of transmission for a trait, it is not as reliable as carrying out crosses that produce large numbers of offspring. Also, pedigree analysis of phenotypes alone cannot be used to develop chromosome maps.

Relatively recent advances in molecular biology techniques, combined with pedigree analysis, make it possible to map genes to chromosomes with a high degree of accuracy. One important technique that is widely used to identify the location of a gene is called **restriction fragment length polymorphism (RFLP) analysis**. This technique is based on the fact that specific breaks or cuts in the nucleotide sequence of a piece of DNA can be made using restriction enzymes. Restriction enzymes cut DNA at specific nucleotide sequences, usually from four to six base pairs in length. For example, the restriction enzyme *Eco*RI recognizes and cuts between the G and A in both strands of DNA containing the sequence GAATTC. A large number of different restriction enzymes with known recognition sequences are readily available commercially. Because, with the exception of identical twins, no two individuals have the exact same nucleotide sequence of DNA in their genome, cutting the DNA from two different individuals with the same enzymes will yield different patterns of variable-length DNA fragments, or RFLPs. An individual's RFLP pattern represents a unique "DNA fingerprint" due to allelic variations in that person's genome. These RFLP patterns are inherited by offspring, and individuals can be assigned a genotype depending on whether they are homozygous or heterozygous for restriction enzyme sites on a particular RFLP.

In humans, RFLPs can be used to test for the presence or absence of a particular allele, especially if the nucleotide sequence of the gene in question is known. If an individual contains a mutation that affects the DNA sequence of a restriction site (either by eliminating a site or by creating a new cutting site), then cutting this DNA with a restriction enzyme will produce different restriction fragments compared with DNA containing the wild-type allele. In this technique, chromosomal DNA is isolated from a tissue sample such as skin, hair, or white blood cells. The DNA is then cut into fragments (RFLPs) with one or more restriction enzymes. These fragments are separated by **gel electrophoresis**. Following electrophoresis, the RFLPs are transferred to a nylon filter that binds the DNA. The filter is then mixed together with a radioactive DNA probe that hybridizes to complementary sequences in RFLPs on the filter. This procedure is called **Southern blotting**. Fragments bound to the probe are visualized by exposing the Southern blot to X-ray film. Exposing the film to the radioactive probe produces dark bands on the film. The exposed film is called an autoradiogram.

What if the nucleotide sequence for a gene is not known? While RFLP analysis is useful for determining genotypes by searching for the presence or absence of a known allele, RFLP analysis must be combined with other techniques when the nucleotide sequence of a gene is not known. How can geneticists find a gene that is responsible for causing a rare genetic disorder? It is possible to look for linkage between the gene causing the disorder and another gene only if other genes close to the mutant gene have already been discovered. Quite frequently, genes near the mutant gene of interest are not known. In these cases, one way to study linkage to map a mutant gene depends on sequences of chromosomal DNA called markers. A marker is typically a sequence that is known to be located at a specific location on a chromosome. Sometime markers include known genes that have been already been identified, but more commonly, markers are short non-protein-coding sequences of DNA. One very valuable group of human chromosome markers are called microsatellites. Microsatellites are usually tandem repeats of dinucleotides (e.g., CACACA) that are highly repeated from individual to individual. Because they are scattered throughout the genome, microsatellites are excellent markers for determining and mapping how close an unknown mutant gene may be to a microsatellite marker on many different human chromosomes. A marker can be detected by hybridizing short pieces of radioactive DNA, called probes, that are complementary to the nucleotide sequence of the marker.

RFLP analysis using markers has proven to be a very powerful technique for mapping chromosomes. Piecing together overlapping RFLPs and other markers has enabled biologists to develop extensive linkage maps of chromosomes in the human genome. In excess of 5000 markers representing different loci in the genome have been identified and are available for use by scientists who are searching for genes that cause inherited disorders in humans. With so many markers available, radioactive probes for many different markers can be hybridized to RFLP fragments from individuals who have a certain disease. RFLP patterns from many individuals with a disease can be compared with each other and compared against RFLP patterns for normal, healthy individuals to search for markers that are linked to the same chromosome as the defective gene.

Phenotypic information gathered from pedigrees combined with RFLP analysis provides very powerful tools that can be applied by geneticists and molecular biologists who are attempting to identify and locate mutant alleles that cause human diseases. The gene that causes cystic fibrosis was one of the first genes for a human genetic disorder that was mapped using RFLP techniques. Cystic fibrosis is an autosomal recessive disease in which individuals suffer from severe respiratory problems, among other symptoms. The cystic fibrosis gene, located on chromosome 7, codes for a protein that normally functions to pump chloride ions out of many body cells. Once the chromosome containing the cystic fibrosis gene was identified, it was possible to study recombination between the cystic fibrosis gene and linked markers to pinpoint the exact locus of the cystic fibrosis gene. Refer to Figure 1 on page 17 for an example of RFLP patterns generated for an individual who is heterozygous for an autosomal recessive condition.

Statistical analysis can be applied to the recombination data generated by RFLP studies to determine the probability that a marker and trait are linked. Based on crossing over, recombination data between a marker and a trait can be assigned an LOD (log of odds) score, which is a statistical comparison of the probability that RFLP data is due to linked loci and the probability that RFLP data is due to unlinked loci. A standard LOD score of 3.0 has been established as the threshold value for linkage. A LOD of 3.0 is equal to approximately a 20:1 ratio that favors linkage. A LOD value of 3.0 or higher is good evidence that two loci are linked, and the recombination frequency can then be used assign the trait a map unit position on a chromosome. A LOD score of less than 3.0, with a recombination frequency of approximately 50%, indicates a high probability that the trait and probe are unlinked. These techniques have been used to pinpoint the loci for a number of different single genes that cause disease in humans. In addition to cystic fibrosis, loci for Huntington's disease, Duchenne muscular dystrophy, Alzheimer's disease, and Parkinson's disease, among many others, have been identified. These traits are a few of the 22 traits that you can study in PedigreeLab. And pinpointing the location for a rare trait is the ultimate goal that you will strive for using PedigreeLab!

With PedigreeLab you will choose a gene from a list of mutant genes that cause actual genetic disorders in humans. Most of these disorders are relatively rare. To determine the location of the gene for the trait that you are following in PedigreeLab, you will first study sample pedigrees to develop a hypothesis on the mode of inheritance for the trait. You will then search large family databases of RFLP data that were generated with DNA probes to different markers on human chromosomes. One of your challenges will be to find a probe for a marker that is linked to the gene for the trait that you are studying. Once you have done this, you will analyze RFLP and pedigree data from families that you will choose as ideal families for generating recombination data. You will be looking for recombination between the locus for the marker and the trait that you are studying. This will be a particularly challenging aspect of PedigreeLab because you must consider parents with the proper arrangement of the marker and trait on a set of homologous chromosomes to be able to detect recombination in the offspring.

With PedigreeLab you will count recombinants and nonrecombinants based on RFLP analysis and the genotypes that you have assigned to individuals in a pedigree. Recombination frequency and LOD values are tallied for you. If you see a LOD score of >3.0, then it is likely that the marker you chose and the trait are linked. You can then use the recombination frequency data to assign a position for the trait on the same chromosome as the marker. If, however, your LOD score is <3.0, and recombination frequency is approximately 50%, then the probability is high that the trait and marker are unlinked and you'll need to start your gene hunting again with a different probe!

References
1. Cummings, M. *Human Heredity*, 3rd ed. St. Paul, Minn: West, 1994.

2. Lander, E., and Botstein, D. "Mapping Mendelian factors underlying quantitative traits using RFLP linkage maps." *Genetics* 121 (1989).

3. Lewin, B. *Genes VI*. New York: Oxford University Press, 1997.

Introduction

In FlyLab you studied modes of inheritance and chromosome maps in *Drosophila melanogaster*. In this laboratory you will use pedigree analysis to simulate the inheritance of genes for human genetic disorders and RFLP analysis to study recombination in humans. Combining RFLP analysis with pedigree analysis will enable you to learn about, and understand, how the location of a gene can be assigned to a chromosome. You will also pinpoint the location of a gene on a chromosome by using recombination data to generate a genetic map of that chromosome.

For your ease in completing each assignment, the background text relevant to the experiment that you will perform is *italicized*, instructions for each assignment are indicated by plain text, and questions or activities that you will be asked to provide answers for are indicated by **bold text**.

Objectives

The purpose of this laboratory is to:
- Demonstrate how pedigree analysis can be used to determine the mode of inheritance for a genetic disorder across several generations of humans.
- Demonstrate how recombination data from genetic crosses can be interpreted and applied to discover the location of a gene on a chromosome and to develop chromosome maps.
- Simulate the use of restriction fragment length polymorphism (RFLP) analysis to develop genetic maps of a chromosome.

Before You Begin: Prerequisites

Before beginning PedigreeLab you should be familiar with the following concepts:
- Chromosome structure and the stages of gamete formation by meiosis (see Campbell, N. A., Reece, Reece, J. B., and Mitchell, L. G., *Biology* 5/e, and Campbell, N. A., and Reece J. B., *Biology* 6/e, chapters 13–15).
- Using FlyLab to study basic principles of Mendelian genetics.
- Inheritance of normal traits and genetic diseases by autosomal recessive, autosomal dominant, and sex-linked inheritance in humans (chapters 14 and 15).
- Use of genetic crosses to develop chromosome maps according to linkage groups (chapters 14 and 15).

Assignments

Twenty-two different mutations of human genes that produce actual genetic diseases are included in PedigreeLab. Some of these mutations produce diseases that you have probably heard about and already studied; in other cases, rarer genetic mutations and diseases are presented. You will use PedigreeLab to study a number of different pedigrees to develop a hypothesis to explain the mode of inheritance for the trait that you selected. Once you have done this, your goal is to study recombination between the mutant gene and markers that you will select, using RFLP analysis to help you determine the chromosome containing the mutant gene, and the location of the mutated gene relative to genetic markers on that chromosome.

For your ease in completing each assignment, the background text relevant to the experiment that you will perform is *italicized*, instructions for each assignment are indicated by plain text, and questions or activities that you will be asked to provide answers for are indicated by **bold text**.

The following assignment is designed to help you become familiar with the operation of PedigreeLab.

Assignment 1: Getting to Know PedigreeLab: Inheritance of Amyotrophic Lateral Sclerosis (ALS)

The first screen that appears in PedigreeLab is the Information view, beginning with Alzheimer's disease.

Click on the popup menu for the mutation list. A list of abbreviations for gene mutations of various human genetic disorders appears. Scroll down this list and view the different mutations and diseases presented. Note that for each mutation you are provided with a short narrative of background information describing the history, phenotypes, and other symptoms created by the mutation.

You are not provided with any clues, however, about the location of the gene for this condition. It is your job to hunt for the chromosomal location of each mutation that you choose!

Choose the ALS abbreviation for the mutated gene that results in amyotrophic lateral sclerosis, also commonly known as Lou Gehrig's disease, named after the famous New York Yankees baseball player who died from the condition. Read the information panel to learn more about ALS.

You may have studied ALS before, and you may also know that another former Yankee, Jim "Catfish" Hunter, was recently diagnosed with ALS. This first assignment is designed to introduce you to the process and intricacies of using PedigreeLab by studying the inheritance of ALS and mapping the locus of the ALS gene to its correct chromosome. The other assignments in this laboratory are based on your ability to use your understanding of genetics, and the basic functions of PedigreeLab that you will learn in this first assignment, to map the mutations in PedigreeLab.

50

1. <u>Determining the Mode of Inheritance for a Gene by Studying Full Pedigrees</u>
   To determine the pattern of inheritance for a particular mutation, click on the Full Pedigree tab at the top of the screen. The pedigree that is displayed on this page is one of 100 sample pedigrees for ALS available in PedigreeLab. To choose a different pedigree, click on one of the up or down arrows in the upper right corner of the screen. The purpose of this Full Pedigree function is to provide you will a large number of different pedigrees that you can evaluate to develop a hypothesis for whether a particular mutation is inherited by autosomal recessive, autosomal dominant, or sex-linked inheritance.

   a. When studying these pedigrees, remember that circles represent females and squares represent males. Parents are connected by a red horizontal line. Black vertical lines from the parents indicate the offspring produced by these parents. Sibs are connected to each other by a horizontal black-lined bracket shown above the symbols. The symbol for an individual that expresses the mutant phenotypes is shaded blue. First generation parents are shown at the top of the pedigree, followed by the second generation, third generation, and so on.

   b. You may already know that ALS is inherited as an autosomal dominant condition. Heterozygotes develop the disease while homozygous recessive individuals are phenotypically normal. Homozygous dominant individuals rarely live to adulthood because of the lethality of this mutation. Study several of the full pedigrees until you are comfortable with the inheritance of ALS as an autosomal dominant trait.

   c. <u>Assigning Genotypes to the Individuals in a Pedigree</u>
      Select a simple pedigree with a first-generation parent that has ALS and a first-generation parent who is phenotypically normal. You can assign genotypes to each of the members of a pedigree as follows:

      (1) To assign a genotype to each parent, double-click on the circle for the first-generation mother. A new box should open prompting you to type in the mother's genotype. Type in "Aa" or "aa" depending on whether the mother is a heterozygote with ALS or a phenotypically normal homozygote, then click OK to close the genotype box. Double-click on the square for the first-generation father and type in the appropriate genotype for the father, then click OK to close the genotype box. The genotype for each parent should appear below each symbol.

      (2) Repeat this process to assign genotypes for the second- and third-generation individuals in the pedigree. Once you have filled in the genotypes for a pedigree, the labeled pedigree can be exported as a GIF file by clicking on the Export Pedigree button at the lower right corner of the screen. A new browser window with your labeled pedigree will appear. You can now print your pedigree from this

window. Analyze at least two other pedigrees to confirm your genotype assignments.

2. <u>Mapping the Locus for a Mutant Gene by Studying Large Family Pedigrees, RFLP Analysis, and Recombination</u>

Identifying the chromosome that contains the mutation that you are studying and mapping the locus for this mutation using PedigreeLab involves your ability to apply what you learned about the mode of inheritance for the mutation, and your ability interpret the results of RFLP and recombination data. This process is part of the challenge and fun of PedigreeLab.

Using the Large Family function is the key to the gene-mapping functions of PedigreeLab. This function simulates databases that present family pedigrees for individuals with different genetic disorders that have been subjected to RFLP analysis using one of 17 different probes for markers on human chromosomes. The pedigrees are arranged according to genotypes for the RFLPs; phenotypes are also shown with these pedigrees. The probes in PedigreeLab do not correspond to actual marker loci in human chromosomes.

You will pick an appropriate pedigree that will allow you to look for recombination between the marker sequence of DNA recognized by the probe and the mutation. Based on recombination frequencies that you generate, and LOD scores, you will then decide whether the probe and mutation are linked to the same chromosome. If they are linked, you will use the Chromosome View function to pinpoint the locus for your trait based on recombination frequencies. All of the traits presented in PedigreeLab will map to their actual chromosomes as found in real life. No shortcuts or hints are given to determine linkage groups; hence, you must be a gene hunter using trial and error and analysis of your data to eliminate loci for unlinked probes and find linked probes that will help you pinpoint the locus for the mutant gene!

a. Begin by clicking on the Large Family tab at the top of the screen. Before you view the family pedigrees, answer the three questions on the expression of the mutant phenotype in a pedigree. You have the option of selecting whether the mutant trait is expressed in grandparents, parents, and offspring. The purpose of these three questions is to better define and narrow your search of the Large Family databases to select for those pedigrees that will be most helpful to study recombination.

(1) *Remember that for most traits you will want to define the search to look for pedigrees that have one parent who is a double heterozygote for the trait and probe (marker) and one parent who is a double homozygote for the trait and probe.*

(2) For this ALS assignment, answer yes to each question. For the other assignments, think carefully about the mode of inheritance of the trait that you are working on before answering yes or no to each of the three questions. Before searching the database, note

that the default probe selected should be the "asr" probe. Click on the Search Database button. Your search of the database will reveal different family pedigrees that are assigned genotypes resulting from RFLP analysis with the asr probe.

b. RFLP Analysis
The first screen that appears after searching the Large Family database displays a pedigree in which individuals are labeled according to their phenotype expressed and assigned a genotype according to the results of RFLP analysis with the asr probe.

    (1)   To interpret each phenotype, note that the symbols for individuals expressing the mutant phenotype are shaded blue. You can assign a genotype for the trait to each individual in the pedigree by double clicking on each symbol as you did with the Full Pedigrees and entering the genotypes for each individual in the pedigree.

    (2)   To interpret the results of the RFLP analysis, genotypes are labeled in each symbol of the pedigree according to the presence (+) or absence (−) of a restriction site in the fragments of DNA detected by the probe. Individuals who are homozygous for the presence of a restriction site are labeled as "+/+." Heterozygous individuals are labeled as "+/−," indicated the absence of a restriction site on one homologue and the presence of a restriction site on the other homologue. Homozygous individuals lacking a restriction site on both homologues of a pair are labeled as "−/−."

    (3)   Notice that an autoradiogram showing the RFLP pattern that would appear for different individuals depending on their RFLP genotype is shown on the right side of the screen. Reading the autoradiogram from left to right, the first lane, labeled "Ref," consists of a series of DNA fragments of known size in kilobases (kb) that are used as size standards, or references for determining the size (in base pairs) of the RFLP fragments. Clicking on the symbol for any individual in the pedigree will show their RFLP pattern in the "selection" lane of the autoradiogram. Compare the size of each DNA fragment with the size standards; note that large DNA fragments are located at the top of the autoradiogram while smaller fragments migrate relatively faster toward the bottom of the autoradiogram. You may want to refer back to Figure 1 in the background section of this lab to help you interpret the RFLP autoradiograms.

c. Choosing a Large Family Pedigree to Study Recombination Between the Trait and Marker
Once you are comfortable with how a Full Pedigree is shown, use the arrows in the upper right corner of the screen to find a pedigree that is

appropriate for studying recombination. Begin by looking for a pedigree with one parent who is a double heterozygote (A/a, +/–) for the trait and probe, and a parent who is a double homozygote for the trait and probe (a/a, +/+). Look carefully at the phenotype and RFLP genotype of the grandparents before assigning genotypes to the parents. Once you have found an appropriate pedigree, you may find it helpful to label all of the individuals in the pedigree before attempting to count recombinant and nonrecombinant offspring. *Hint*: If you are having difficulty finding the correct pedigree, use the Genetic Calculator function described below.

(1) The **Genetic Calculator** function can be very useful for helping you determine which pedigrees will be best suited for studying recombination between the trait and marker. This function allows you to specify genotypes in the parents of a full pedigree that show the trait and marker linked to a pair of homologous chromosomes or unlinked on different chromosomes.

To use this function, click on the Genetic Calculator tab at the top of the screen, then click on the popup arrow next to the box labeled "Trait & probe autosomal, linked" at the top of this window. This popup menu allows you to design genotypes for different individuals depending on whether you think that the trait you are trying to map is an autosomal or sex-linked trait, and whether or not this trait is linked or unlinked to the probe for a given marker. Because you know from studying the full pedigrees that ALS is inherited as an autosomal dominant condition, select "Trait & probe autosomal, linked" to design genotypes that would show the ALS trait and asr marker as linked on an autosome.

Notice that on the far left of this box there are four popup menu boxes that you can use to select the trait and marker combination for each homologue of a pair in a female parent. Click on one of the upper two boxes: These allow you to choose the genotype for the trait. Note that the dominant allele for ALS is shown in capital letters and the recessive allele for ALS is shown in lowercase letters. Click on one of the lower two boxes: These allow you to choose the genotype for the probe (marker). Note that the genotype for a homologue with a restriction site is shown as "asr+," and the genotype for a homologue lacking a restriction site is shown as "asr–."

It is important that you determine the correct linkage combination for the parent who is a double heterozygote (A/a, +/–) for the trait and restriction site. Do this for the female parent. For the first homologue, select a linkage combination with "ALS" and "asr+" on the first homologue. Select the opposite homologue with the linkage configuration of "als" and "asr–."

Repeat this process for a male parent who is a double homozygote for the trait and restriction site (a/a, +/+). For this parent, the linkage combination is the same on both homologues.

Once you have selected the genotypes for each parent, click on the Calculate Punnett Squares button. You should now see Punnett squares that show nonrecombinant and recombinant genotypes that would result from a cross between the parents with the trait and restriction site arrangements that you just set up. You are now ready to return to the Large Family pedigree function to look for a pedigree with parents that show the genotypes for the trait and marker combination that you set up with the Genetic Calculator. Once you have done this, use the results of the Punnett squares to help you look for the genotypes that you will count as recombinants and that you will count as nonrecombinants. *Hint*: To help you as you search the Large Family pedigrees, you may want to export the results of your Punnett squares from the Genetic Calculator and print a copy of these results to have in front of you as you look at the large family pedigrees to count recombinants and nonrecombinants.

(2) Counting Recombinants and Nonrecombinants
Once you have located a Large Family pedigree with the double heterozygous and double homozygous parents that you determined with the Genetic Calculator function, begin to tally recombinant and nonrecombinant offspring by clicking on the arrow buttons at the bottom of the screen.

After counting offspring for one pedigree, choose another useful pedigree and look for recombination. Note that as you tally recombinants and nonrecombinants, a running tally for the total number of offspring that you have counted is shown at the bottom of the screen along with recombination frequency and LOD score. You will need to count offspring from several pedigrees to accumulate data that will help you determine if the trait and probe are linked. Remember that you are looking for a LOD score of 3 or greater, and a recombination frequency of less than 50% as good evidence that the trait and probe are linked. If you are seeing evidence for linkage, then you are ready to map the trait to a chromosome.

Conversely, a recombination frequency of 50% indicates that the probe and trait are probably unlinked. This can usually be determined after analyzing just a few pedigrees. If this is the case, choose another probe and try again! You should find that recombination data using the asr probe indicates that the asr marker is not linked to the ALS gene. Click on the Chromosome

View and note that the asr probe is located on chromosome 1; therefore, you can probably rule out this chromosome as the location of the ALS gene. You may also want to rule out using the other probes on this chromosome in your next search. You may find that you will have to search for recombination with several probes before you find one that is linked to your trait.

To look for linkage with another probe, return to the Large Family function. Click on the New Search button. This time, search the database for pedigrees using the "juva" probe. Once you have done this, count recombinant and nonrecombinant offspring again. This time you should see evidence for linkage of the juva probe and ALS gene. Follow the next procedure to map the ALS gene to the correct chromosome.

d. Mapping the Mutation to a Chromosome
Now that you have evidence for linkage, click on the Chromosome View tab at the top of the screen. Genetic maps appear that show loci for markers on five different chromosomes. Notice that the juva marker is located on chromosome 21. To position the ALS gene, click and drag the ALS arrow on the upper right side of this window. Drag the ALS arrow to chromosome 21 and position the arrow relative to the juva marker according to the recombination frequency value that you generated with the Large Family pedigrees. For example, if you generated a recombination frequency of 15%, you should move the ALS arrow to a position that is 15 map units (centimorgans) away from the juva locus. You will not know, however, on which side of the juva locus to position the ALS arrow until you study linkage between the ALS locus and the other markers on chromosome 21.

Notice the other markers on chromosome 21, "ddy" and "utc3." Return to the Large Family function and repeat a database search using probes for each of these markers to better pinpoint the location of the ALS gene. The more offspring you tally, the more reliable will be your estimates of the recombination frequencies. Counting 50–100 offspring for each probe will give you good values. You can also refine your estimate by averaging the map locations for the different probes. You should find that the ALS gene maps to a position just above the juva marker. See your instructor for the exact location of the ALS gene.

e. Printing Your Chromosome Map and Recombination Data
Print a copy of the chromosome map that you developed by clicking on the Export Graphic button in the lower right corner of the screen. A new GIF image window will open which you can now use to print your map.

To print your recombination data, return to the Large Family view. You can export a single pedigree that you have labeled by clicking on the Export Pedigree button and printing from the GIF image window that appears. You can also export all of the data (total number of recombinants, nonrecombinants, recombination frequency, and LOD scores) for each probe that you used by clicking on the New Search button, then clicking on the Export Whole List button. With this function you will open a window with your data that functions as a lab notebook. You can add any comments that you'd like to your data and print a copy of this data as evidence of your gene-hunting progress.

Assignment 2: On Your Own: Mapping Other Mutations

Now that you have experienced how PedigreeLab functions to map the location of the ALS gene, you are on your own to be a gene hunter for some of the other mutations in PedigreeLab!

1. Follow the guidelines in the ALS assignment to help you identify the location and approximate locus for each of the following mutations:

   - Huntington's disease (HD)
   - Duchenne muscular dystrophy (DMD)
   - Werner's syndrome (WRN)
   - Cystic fibrosis (CF)
   - Two additional mutations of your choice.

2. For each mutation, perform the following:

   a. **Study several Full Pedigrees, then develop a hypothesis to explain the mode of inheritance for the mutation.**

   b. **Use the Large Family pedigree function to identify pedigrees with parents that will produce offspring that you can use to look for recombination. Think carefully about the mode of inheritance for the trait, then decide whether to answer yes or no to the three questions that ask you to define whether the trait is expressed in grandparents, parents, and offspring before searching the Large Family database.**

   Use the Genetic Calculator to help you select the correct Large Family pedigrees and decide what genotypes to look for when counting recombinant and nonrecombinant offspring. Once you have selected the correct pedigree, count the number of recombinant and nonrecombinant offspring. Do this until the recombination and LOD data indicate whether the trait and probe are linked or unlinked. If the trait and probe are linked, use the Chromosome View function to assign a position for the mutant gene on the correct chromosome. If the trait and probe are unlinked, then pick another probe to test for recombination and repeat the process until you find a probe that is linked to the trait.

c. **Print your recombination data for any probes that you used to search for linkage.**

d. **Once you have found the chromosomal location for the mutation, be sure to test all of the other probes on that chromosome that are linked to the mutation so you can better refine and pinpoint the location of the trait relative to the markers on the chromosome. Your goal is to develop as accurate a map as you can. Print your chromosome map when you are done.**

e. **Show your recombination data and chromosome maps to your instructor to see how accurate your mapping experiments were.**

Assignment 3: Group Assignments

The Human Genome Project is a worldwide initiative to develop genetic maps for the 22 autosomes and two sex chromosomes in humans. One long-term goal of the Genome Project is to identify and determine the nucleotide sequence for an estimated 100,000 genes that are thought to be present in the approximately three billion nucleotides of DNA that comprise the human genome. The Genome Project began in 1990 and was targeted as a 15-year project with an estimated cost of $3 billion. Although this project originated in the United States, labs around the world are contributing to the Genome Project. Many new technologies for identifying and sequencing genes have been developed as a result of the project. Because of this, the progress of the Genome Project is actually slightly ahead of schedule and just under budget. One technique being used by Genome Project scientists involves using RFLP analysis and other similar techniques to develop extensive genetic maps of chromosomes in humans and other animal and plant species such as the nematode Caenorhabditis elegans; Drosophila melanogaster; the common house mouse, Mus Musculus, and the flowering plant Arabidopsis thaliana. Some labs involved in the Genome Project are working on mapping assigned segments of certain chromosomes. Other groups are involved in checking the accuracy of maps completed to date.

The following exercises are designed to have you investigate the accuracy of genetic mapping as simulated in PedigreeLab, and to enable you to map an entire chromosome. Work together in a group of four students to complete the following assignments.

1. Select two mutant genes that you have not mapped already. Divide your group into pairs, and have each pair of students work together to identify the chromosome and approximate locus for each of the two mutations. Once each pair has mapped the two genes, compare your results with each other. Did each pair in your group identify the correct chromosome(s)? Why or why not? If you cannot agree on the correct chromosome, go back and review your work until everyone in your group can agree on a correct chromosome for each mutation.

Once you have agreed on a chromosome location, compare the accuracy of the locus that you identified for each mutation. How accurate were the maps when you compared them? How many map units did the two maps differ by? Explain possible reasons for any differences in loci assigned on each chromosome mapped.

2. You have been selected as the leading scientist directing a group of scientists responsible for mapping the X chromosome. Pick three other students who can work as your partners to narrow down your hunt to map the X chromosome.

How would you begin? Should you assign each student a mutant gene that you have not yet mapped and have each person study the inheritance for that gene? Or should you work as a group and look for inheritance patterns to first narrow your search for genes that appear to be X-linked? Decide on an approach to use, then work together to develop a map of the X chromosome that will show all of the X-linked mutations present in PedigreeLab. Because PedigreeLab will show different regions of the X chromosome depending on the probes you are using, PedigreeLab will not show one complete full-length map of the X chromosome. Your group will have to piece together the different maps of the X chromosome shown in PedigreeLab to draw one complete map.

Once you have done this, compare your map with that of another group of students in your class that has worked on the X chromosome.

How accurate were the two sets of maps? Were there any differences in genes mapped to the X chromosome when the maps were compared? If there were differences, work together with the other group to resolve these differences until you can agree on an accurate and complete map of the X chromosome. Print your map and show it to your instructor for comparison with the correct map.

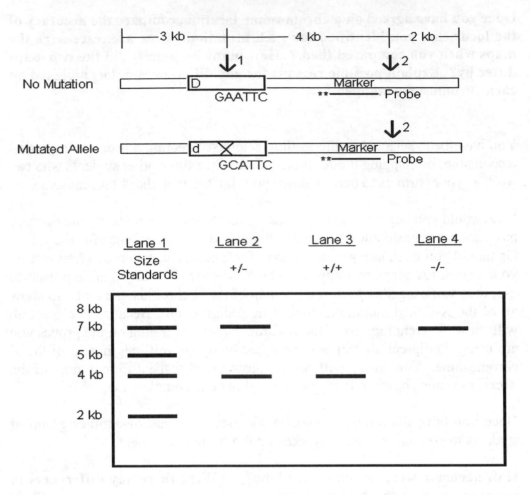

Figure 1: Diagrammatic Representation of Restriction Fragment Length Polymorphisms (RFLPs) for a Recessive Trait. Shown is an example of two homologues from an individual who is heterozygous (Dd) for a recessive gene that is linked to a genetic marker. Arrows indicate the restriction enzyme recognition sequence that would be cut by restriction digestion of each chromosome with the enzyme *Eco*RI. Note that two sites (1, 2) are present on the homologue with the dominant allele (D), while only one site (2) is present on the homologue with the recessive allele (d). The absence of restriction enzyme cutting site number 1 in the recessive allele due to a single nucleotide mutation is indicated with an "X." The location of a radioactive probe that is complementary to the marker sequence is also

60

shown. The pattern of restriction fragment length polymorphisms that would be generated for these homologues is visualized after gel electrophoresis of the digested DNA fragments, transfer of the DNA fragments to a filter, and hybridization of the fragments to the radioactive probe, followed by exposing the filter to X-ray film to produce an autoradiogram. In the autoradiogram, lane 1 shows the migration of DNA size standards of known length that are used to determine the nucleotide length of each RFLP. Lane 2, shows RFLPs for a heterozygote (+/–) where the genotype is designated according to the presence (+) or absence (–) of a restriction site; cutting with *Eco*RI yields two fragments, 7 kb and 4 kb in size. For comparison, lanes 3 and 4 show RFLPs for individuals who are homozygous for the restriction site, +/+ or –/–.

TranslationLab

Background

Genetic information is stored in cells as deoxyribonucleic acid (DNA). In addition to functioning as the hereditary material for living organisms, the information stored in DNA as **genes** is the basis for cell metabolism because all proteins are synthesized from genes. However, DNA is not directly copied into protein. Protein synthesis requires a deciphering of the genetic information stored within DNA whereby the sequence of deoxyribonucleotides in DNA are copied into strands of ribonucleic acid (RNA) during a process called **transcription**. During **translation**, different RNA molecules—specifically messenger RNA (mRNA), transfer RNA (tRNA), and ribosomal RNA (rRNA)—are used to specify the amino acids that are incorporated into a protein. Because it has been well established that this flow of genetic information is universally followed by living cells—including bacteria, yeast, and human cells—this concept is often referred to as the *central dogma of the genetic code*.

The structure of DNA and its role as a genetic material has not always been so clearly understood. In the 1900s, many scientists suggested that proteins were a key component of cell metabolism; however, the association between genes and proteins was not known. During this time, Sir Archibald Garrod and William Bateson observed human patients who demonstrated rare diseases caused by deficiencies in metabolic pathways involving amino acids. Garrod and Bateson followed the patterns of genetic inheritance of these disorders within families and concluded that inherited information controls metabolism in a cell. The terms gene and enzyme were not even used at this time.

In the 1930s, George Beadle and Edward Tatum performed experiments with a bread mold, *Neurospora crassa*, that provided substantial evidence for the relationship between genes and proteins. Beadle and Tatum developed and observed several mutant groups of *Neurospora* that were identified by comparing the ability of these mutants to grow on plates with minimal nutritional medias with that of wild-type *Neurospora*. Beadle and Tatum discovered different mutants that were defective in their ability to synthesize the amino acid arginine. Supplementing these mutants with other amino acids allowed some of these mutants to synthesize arginine. Beadle and Tatum hypothesized that the amino acid supplements that allowed the mutants to synthesize arginine were amino acids that the mutants could not synthesize on their own due to a mutation that caused a loss of enzyme activity. These results led Beadle and Tatum to suggest a *one gene - one enzyme hypothesis*, in which they reasoned that a single gene is important for determining the synthesis of a single enzyme. This hypothesis, however, did not account for how DNA was deciphered by a cell to produce a protein. In the 1940s and early 1950s, several other scientists performed experiments with bacteria and viruses to present strong evidence that genes were made of DNA, but the process by which a protein could be produced from a gene was still unknown.

The discovery of the double-helical structure of DNA by James Watson and Francis Crick in 1953 clearly established that hereditary information in cells is encoded in the nucleotide sequences contained within DNA. Although this landmark discovery represented a significant advance in the history of biological research, a basic question still existed: How was the nucleotide sequence of genes interpreted or decoded by a cell to provide that cell with the instructions for the synthesis of a protein?

A number of different investigators carried out studies to demonstrate that RNA was synthesized from DNA and that RNA performed a central role in protein synthesis. In 1961, Marshall Nirenberg and Heinrich Matthaei used an in vitro cell-free protein-synthesizing system to provide the first evidence for protein-coding sequences of RNA nucleotides. In a cell-free system, organelles such as ribosomes and other factors (including amino acids, tRNAs, an mRNA template, and a number of cofactors required for translation) can be added together to synthesize proteins in a test tube. Extracts of organelles and the molecules required for translation are often isolated by the lysis of bacteria or animal cells that are highly active in protein synthesis. Nirenberg and Matthaei used extracts from bacteria called *Escherichia coli*. The addition of radioactive amino acids to a cell-free extract allows biologists to follow the rate and specific sequences of proteins that are translated in the assay. Because mRNA had only recently been discovered and was not yet easily isolated from cells, Nirenberg and Matthaei synthesized RNA **homopolymers**, single strands of RNA containing only one ribonucleotide in each strand (for example, UUUUUUU, AAAAAAA), by using a bacterial enzyme called polynucleotide phosphorylase. This enzyme does not require a DNA template to synthesize strands of RNA. The homopolymers synthesized by polynucleotide phosphorylase were then added to the cell-free system, and the incorporation of radioactive isotopes into protein was measured. In a cell-free system, translation begins at multiple and random sites along a nucleotide sequence. Keep this in mind when completing the assignments for this laboratory.

Although these early experiments did not determine the number of nucleotides required for a codon, Nirenberg and Matthaei concluded that certain RNA sequences coded for specific amino acids. For example, poly A codes for lysine. Subsequent experiments by Nirenberg and others using RNA heteropolymers, combinations of different ribonucleotides, served to further delineate the assignment of specific nucleotide sequences to individual amino acids. In 1964, Marshall Nirenberg and Philip Leder used an increased understanding of the function of tRNA and the function of ribosomes as RNA-binding organelles to establish that the genetic code is interpreted as three-ribonucleotide sequences, called **codons**, that specify only one amino acid. Nirenberg and Leder's experiments also provided a better understanding of how the anticodon portion of a tRNA molecule interacts with a codon by base pairing during translation.

Gobind Khorana performed similar experiments with a cell-free system to which long sequences of RNA molecules consisting of repeating dinucleotides (for example, ACACAC), trinucleotides, and tetranucleotides were added. The results of Khorana's experiments, and the work of many others, served to identify new codons as well as confirm the specificity of many codons that were previously identified. In particular,

Khorana concluded that certain sequences, such as a triplet contained in polymers of GAUA, function as termination signals because they do not code for the incorporation of an amino acid into a peptide.

It was apparent from many of these studies that the genetic code is degenerate or redundant, because although each codon codes for only one amino acid, most amino acids are specified by more than one codon. The *wobble hypothesis* was proposed by Francis Crick to explain how the first two nucleotides of a codon are more important for tRNA binding to an anticodon than the third nucleotide. Modified base-pair rules that occur with U at the third position (for example, U may pair with A or G at the third position of a codon) suggested a rationale for why the number of different tRNAs inside a cell does not need to equal the number of codons that code for amino acids.

In this laboratory you will have the opportunity to simulate many of the early experiments involving cell-free extracts that were essential for deciphering and determining the genetic code. You will investigate how polyribonucleotide sequences that you create can be translated in a cell-free system to produce sequences of amino acids, and you will interpret the results of your experiments to help you learn how the genetic code is deciphered.

References
1. Alberts, B., et al. *Essential Cell Biology*, 1st ed. New York: Garland Publishing, 1998.

2. Beadle, G. W., and Tatum, E. L. "Genetic Control of Biochemical Reactions in *Neurospora*." *Proceedings of the National Academy of Science*, USA 27 (1941): 499-506.

3. Nirenberg, M. W. "The Genetic Code: II." *Scientific American*, March 1963.

Introduction
TranslationLab will allow you to study the importance of the nucleotide sequence of mRNA as the fundamental basis for the genetic code universally deciphered by living cells. You will produce sequences of ribonucleotides that will be translated into protein to simulate the landmark experiments involving cell-free extracts that were essential for interpreting and understanding the genetic code.

Objectives
The purpose of this laboratory is to:
- Study the relationship between the nucleotide sequence of a mRNA molecule and protein synthesis.
- Simulate pioneering experiments that were used to delineate the genetic code.
- Demonstrate how a mutation in the nucleotide sequence of a mRNA molecule results in a change in the amino acid sequence of a protein.

Before You Begin: Prerequisites

Before beginning TranslationLab you should be familiar with the following concepts:

- The structure of a polypeptide (see Campbell, N. A., Reece, J. B., and Mitchell, L. G., *Biology* 5/e, and Campbell, N. A., and Reece J. B., *Biology* 6/e, chapter 5).
- The structure and functions of messenger RNA (mRNA), ribosomal RNA (rRNA), and transfer RNA (tRNA) (chapter 17).
- The flow of genetic information in a cell; the major processes involved in transcription and translation (chapter 17).
- Describe how point mutations in a gene can affect the amino acid sequence of a protein (chapters 5, 14, 17).

Assignments

During the late 1950s and early 1960s researchers were able to solve one of the major secrets of life: how genes worked. The problem the researchers were trying to solve was how a linear sequence of four nucleotides (A, G, C, and U) determined the amino acid sequence of proteins, which were made out of up to 20 different amino acids. The following assignments are designed to help you reproduce some of the experiments that the scientists used to figure out how this was accomplished. Your mission, should you choose to accept it, is to crack the genetic code of life.

For your ease in completing each assignment, the background text relevant to the experiment that you will perform is *italicized*, instructions for each assignment are indicated by plain text, and questions or activities that you will be asked to provide answers for are indicated by **bold text**.

The following assignment is designed to help you become familiar with the operation of TranslationLab.

Assignment 1: Getting to Know Translation Lab: The Genetic Code

A major step forward in figuring out the code was the discovery by Nirenberg in 1961 that a cell-free extract made from E. coli cells could translate RNA added to the extract into proteins. The composition of the newly synthesized proteins could be determined by measuring the incorporation of radioactive amino acids into these proteins as they were translated. In his first experiment he made poly U RNA, using the enzyme polynucleotide phosphorylase, and translated it into a peptide of polyphenylalanine using the cell-free extract. This was definitive proof that RNA could code for the synthesis of proteins and gave the first possible assignment of a nucleotide code to the amino acid it specified.

1. *Because having each nucleotide code for only one amino acid would allow for only four different amino acids to be incorporated into a protein, it was obvious to researchers that there had to be a conversion between multiple bases and each amino acid.*

   Would two nucleotides at a time be sufficient to provide enough codons to code for all 20 amino acids? Why or why not? How many amino acids could be coded for by codons containing only two nucleotides? Will three nucleotides per codon work? Why or why not? Explain your answers.

To answer these questions using TranslationLab, click the Start Experiment button on the input screen of TranslationLab. For each of the four bottles of ribonucleotides that appear, click on the arrow to select a nucleotide. Do this for two nucleotides initially. Click the Make RNA button to display the sequence of mRNA that you created. Click Add to Notes to create a record of your experiment. To translate this sequence into amino acids, click on the To Translation Mix button. Click Add to Notes to add the amino acid sequence to your notebook. Continue this process until you are able to answer the questions above.

2. *Once it was determined that codons consisted of three-nucleotide sequences, the specificity of each codon could be determined.* Use TranslationLab to determine what poly U codes for by performing the following exercise. Click the Start Experiment button on the input screen of TranslationLab. For each of the four bottles of ribonucleotides that appear, click on the arrow to select the uracil (U) nucleotide. Click the Make RNA button to display the poly U sequence of mRNA that you synthesized. Click Add to Notes to create a record of your experiment. To translate this sequence into amino acids, click on the To Translation Mix button. **What does poly U code for?** Click Add to Notes to add this peptide sequence to the poly U sequence. Repeat the same procedure to make polynucleotides of each of the other three nucleotides. **What amino acids do these polynucleotides code for?** Refer to a codon chart (see Campbell, N. A., Reece, J. B., and Mitchell, L. G., *Biology* 5/e, and Campbell, N. A., and Reece J. B., *Biology* 6/e, chapter 17). **Are the amino acids coded for by the polynucleotides you created consistent with what you would expect based on the codon chart?**

3. *Although the Nirenberg experiments showed that RNA did determine the amino acids in the protein, they did not show how many bases were used for each codon, whether the codons were overlapping (is the second codon read from the second base of the first codon [overlapping] or from the first base after the last base of the first codon? [no overlap]), or whether there could be bases in between the codons that did not code for anything (AUCGGGAACGGGACAGGGG, for instance, where the G's in between AUC, AAC, and ACA aren't translated into amino acids—just as we use spaces to separate words in a sentence). Khorana developed a means to produce polydinucleotide and later, polytrinucleotide and polytetranucleotide sequences of DNA that could then be transcribed into RNA to be added to the cell-free translation mix.*

If the code is read two bases at a time, what result would you expect for a polydinucleotide such as AUAUAUAU? Try it and see whether your prediction was correct. From your results can you say whether the code is even or odd? Will you get a different result with UAUAUAUA than you did with AUAUAUAU? This result shows that in these crude extracts translation starts at a random location in the RNA sequence. Translate all possible dinucleotides with TranslationLab. Did you get all of the amino

acids? If not, which ones are missing? Did you get any amino acids more than once? Which ones? What does this tell you about the code?

4. From what you have already discovered, what do you think will happen if you use a polytrinucleotide such as AAC?

   Try it. To help analyze your results, once you have entered the polytrinucleotide sequence, click the Make RNA button and then add this sequence to your notebook by clicking the Add to Notes button. Add the peptide sequence produced from this RNA to your notebook as well.

   Did you get the result you expected? Explain what happened.

   Will ACA or CAA give a different result? From these results can you now tell how many bases there are in a codon? If so, how many are there and how do you know this? Comparing this result with the result from polydinucleotide AC, can you now specify a codon for one of the amino acids incorporated by these templates? If so, which codon and which amino acid go together? By elimination, can you assign another codon–amino acid pair? (*Hint*: Using what you know now, look back at the dinucleotide experiment with AC.) What is it? Try CAC next. Did the results support your codon assignment? Is there evidence here that one of the amino acids must have more than one codon that codes for it? If so, which one? Confirm your results by referring to a codon chart (see Campbell, N. A., Reece, J. B., and Mitchell, L. G., *Biology* 5/e, and Campbell, N. A., Reece, J. B., *Biology* 6/e chapter 17).

5. What do you think will happen if you translate a tetranucleotide? Try translating the tetranucleotide CAAG. Did you get the result you expected? Can you now assign a codon to any of the other amino acids that appeared in problem 5? (Don't worry about any new amino acids that showed up here, just solve the codons for the amino acids in problem 5.) If so what are they? Test your assignment with AACG. Did this confirm your results? Using the above data and any other experiments that are necessary, assign amino acids to all possible codons that do not include G or U, only various combinations of A and C.

6. Now try AU, AAU, and AUU. Did you notice something different this time? What happened, and how would you explain this unusual result? List any new codon assignments that you were able to make from these experiments. Use tetranucleotides to figure out which amino acids go with the codons that can be produced using only A and U. What unusual result did you see with some of the tetranucleotides and what is your explanation for this result?

7. **Now try GGG, GGA, GGC, and GGU. What amino acid showed up in all four experiments? Are there any codons shared in common by these four reactions? If not, then what must be true to explain your results? Can you propose a codon or codons for the amino acid that showed up in all four experiments? Do the codons that you've just assigned to this amino acid have anything in common? What is it? Use tetranucleotides to prove that your assignment is correct. Comparing these results with the ones above, can you say whether some positions in the codon are less important than others in specifying which amino acid is coded for?**

Assignment 2: Altering the Genetic Code: Mutations

Single nucleotide changes (point mutations) in the sequence of a gene can result in changes in the amino acid sequence of a protein produced from the mutated gene. One of the most well studied examples of the effects of a mutation on the sequence of a protein involves the oxygen-transporting protein hemoglobin. A mutation creates an altered form of hemoglobin that produces the genetic disorder called sickle-cell disease (sickle-cell anemia). You will learn more about hemoglobin and the effects of mutations in the hemoglobin gene in HemoglobinLab. The purpose of the following assignment is to demonstrate the effect of a point mutation on the amino acid sequence of a protein.

1. *Sickle-cell disease results from a point mutation in the second nucleotide of the codon GAA, which results in a change in the amino acid at position 6 in the hemoglobin protein.*

   Synthesize a mRNA from the trinucleotide sequence GAA. Enter this sequence in your notebook. Translate this mRNA and enter the results in your notebook. Synthesize and translate the trinucleotide GUA and do the same for the trinucleotide GAG. Assign codons to each amino acid produced from the three mRNA sequences. (*Hint*: Consider what you know about the sequences for stop codons from attempting to assign codons for each amino acid.)

   What amino acid does the codon GAA specify? Which amino acid is incorporated into the sickle-cell hemoglobin molecule when this codon is mutated to GUA? Perform other experiments if necessary to confirm your codon assignments to answer this question.

Assignment 3: Group Assignment

Because the genetic code is a universal code in biology, in general the nucleotide sequence of important genes is highly conserved across many different species of organisms. It is very common for 70% or more of the nucleotides in a gene to be conserved among very different organisms. Redundancy in the genetic code allows for small differences in the nucleotide sequence for a given gene without significant variations in the amino acid sequence of a protein. For example, the nucleotide sequence of the gene for insulin, the peptide hormone required for glucose uptake by many body cells, is well conserved (greater than 80% similarity) in many vertebrate species. As a result of this conservation of nucleotide sequence, comparing the peptide sequence for insulin from cows, humans, sheep, dogs, and rats often shows fewer than six or seven differences in amino acid sequence. Point mutations, as well as insertion and deletion mutations can create changes in a gene that dramatically alter the protein produced. Changing the genetic code changes the protein that is translated. To help you understand why the nucleotide sequences for important genes are highly conserved, work together in a group of four or five students to complete the following assignment.

Imagine that you have just purified a new protein from the brain of adolescent males that you believe may be responsible for excessive hair-combing behavior. From peptide-sequencing experiments, you have determined that this protein contains repeats of the following peptide sequence: Trp-Met-Asp-Gly. Determining the nucleotide sequence of mRNA that was used to translate this part of the protein will enable you to identify the chromosomal location of this new gene and allow you to isolate and clone this gene. It is known that this peptide sequence is highly conserved among males that demonstrate excessive hair-combing behavior which suggests that this portion of the protein is important for its functions. In addition, a mutant form of this protein has also been discovered that appears to result in the loss of the excessive hair-combing behavior. This mutant sequence arises from a single point mutation in the nucleotide sequence of the normal (wild-type) gene for this protein that creates alterations in the amino acid sequence of the protein produced. The peptide sequence from this mutant protein is Met-Tyr-Val-Cys. Use TranslationLab to complete the following exercises.

1. Determine the sequence of a mRNA that could be used to translate this peptide.

2. Can you determine another sequence of mRNA that would also code for this peptide? Why or why not? Explain your results.

3. Once you have deciphered the mRNA sequence for the normal protein, introduce changes in this sequence until you have determined the nucleotide sequence that specifies the mutated peptide sequence. Examine this mRNA sequence and identify the codon or codon(s) that were altered to create the mutant peptide.

69

HemoglobinLab

Background

Virtually every chemical reaction and activity that occurs in a living cell requires proteins. A multitude of different types of proteins perform a wide range of functions that include roles in cell support and shape, cell motility, cell communication, protection against foreign materials, cell reception, cell adhesion, catalytic functions as enzymes, and the transport of molecules. This great diversity of protein functions is a direct result of the structural organization and structural properties of proteins.

The three-dimensional conformation of a protein, also known as **protein structure**, is determined by the arrangement of amino acids that are held together by peptide bonds to form a polypeptide (20 or more amino acids linked together by peptide bonds). The specific sequence or order of amino acids in a polypeptide is known as the **primary structure** of a protein. In many proteins, chemical bonding between amino acids produces proteins with higher-order arrangements known as **secondary** and **tertiary structure**. In addition, for certain proteins—particularly enzymes, structural proteins, and transport proteins—a complete and functional protein consists of multiple polypeptide chains (subunits) that must wrap around each other in an arrangement known as **quaternary structure**. The overall conformation or structure of a given protein provides that protein with the unique structural characteristics that are necessary for the proper functions of that protein. Hence, disruption of protein structure, a process known as protein denaturation, drastically alters the functions of a protein.

One of the most extensively studied examples of the relationship between protein structure and function involves **hemoglobin**, the oxygen-transporting protein in human red blood cells. Adult human red blood cells contain relatively few organelles compared with other body cells. In the most basic sense, human red blood cells are essentially membrane sacs filled with hemoglobin. On average, a single red blood cell contains approximately 250 million hemoglobin molecules! The abundance of hemoglobin in red blood cells and the unique structure of the hemoglobin molecule itself accomplish the primary function of red blood cells: to transport oxygen from the lungs to body cells, tissues, and organs.

In addition to transporting oxygen, the conformation of hemoglobin contributes to the biconcave disk shape of human red blood cells. The shape of these cells provides them with the necessary flexibility to flow through thin-diameter blood vessels, such as capillaries, with a minimal amount of friction.

A single hemoglobin molecule consists of four subunits of a polypeptide known as globin. Globin polypeptides are synthesized from a large family of genes that are highly conserved among many species of vertebrate and invertebrate organisms. This family includes relatives such as myoglobin, an oxygen-storage protein present in most vertebrates. Several different globin polypeptides are used to transport oxygen in red blood cells, including some that are used only during fetal development. Adult human red blood cells contain hemoglobin molecules, which involve two alpha globin (α-globin) subunits and two beta globin (β-globin) subunits that wrap around each other. In the center of each globin subunit is a single iron-containing organic ring known as the heme group. One oxygen molecule can bind to the iron atom in each heme group; therefore, each hemoglobin molecule can bind to and transport a maximum of four molecules of oxygen.

The oxygen-carrying capability of hemoglobin depends on the electron configuration of the iron atom. Within the heme group, the iron atom exists as a transition metal in a divalent state called ferrous iron (Fe^{+2}). The charged nature of an iron ion is essential for oxygen to bind to the heme group. In addition, oxygen binding to the heme group requires a change in the oxidation state of iron that allows the iron to bind oxygen without oxidizing the oxygen or the iron atom itself. This is accomplished by complexing the iron to four nitrogen atoms in the porphyrin ring and one amino acid in the globin subunit, and surrounding the ring with a cluster of hydrophobic amino acids in each globin subunit, thereby creating a hydrophobic pocket around the heme group. This configuration holds the iron atom at a displaced position above the plane of the heme group, which is the ideal conformation for binding and holding oxygen.

In addition, hemoglobin is an efficient carrier of oxygen because the globin chains exhibit an interaction known as cooperativity. In cooperativity, binding of a single oxygen molecule to one heme group results in an increased binding affinity for oxygen to the three other heme groups. Cooperativity occurs because the binding of one oxygen molecule to one heme group creates a shift in the conformation of each globin chain that produces a change in the overall quaternary structure of the entire hemoglobin molecule. This transition in conformation creates a molecule that favors the binding of additional oxygen molecules. The conformation change that occurs during cooperativity is due to the bonds or contacts of specific amino acids that connect the alpha and beta chains to each other and allow these chains to interact with each other.

In 1949, Linus Pauling provided biologists with a significant insight into the molecular basis for **sickle-cell disease** (sickle-cell anemia) by using **gel electrophoresis** to demonstrate that hemoglobin molecules isolated from normal patients and from those with sickle-cell disease differed in their rate of migration. This observation led Pauling to suggest that a difference in amino acid sequence accounted for the migration differences of these proteins. Using peptide sequencing techniques, Vernon Ingram subsequently demonstrated that the differences between normal hemoglobin and sickle-cell hemoglobin are due to an amino acid difference in the primary structure of the two proteins. This amino acid difference occurs because of a point mutation in one of the globin genes.

The most common mutation in a globin gene of an individual with sickle-cell disease involves a substitution in the codon that codes for the amino acid glutamic acid at position 6 in the β-globin polypeptide. Recall that single-nucleotide changes in the DNA sequence of a gene are known as **point mutations**. Point mutations in a gene sequence can result in the synthesis (transcription) of messenger RNA (mRNA) molecules with an altered base sequence. Depending on the location of a mutation within a codon, a mutation may or may not affect the protein coded for by a particular mRNA. The altered codon in sickle-cell disease results in the substitution of valine for glutamic acid at position 6. This disruption in the primary structure of the globin polypeptide occurs in a location on the globin subunit that is necessary for the proper folding of hemoglobin into its three-dimensional conformation that is essential for it to function as an oxygen-transport protein. As a direct result of this change in hemoglobin structure, sickle-cell hemoglobin binds oxygen with a much lower affinity than normal hemoglobin. In addition, red blood cells in the sickle-cell patient lose their characteristic biconcave disk shape and assume an irregular, elongated sickled configuration that greatly diminishes their movement through blood vessels. Sickled red blood cells frequently clump together and block blood flow through the capillaries. Because of changes both in hemoglobin and red blood cell structure, the sickle-cell patient suffers from decreased oxygen delivery to body organs and a variety of other painful conditions related to poor circulation of red blood cells and inadequate oxygen content within the body. The study of hemoglobin biochemistry and sickle-cell disease has provided biologists with an invaluable understanding of the molecular basis for disease.

In this laboratory you will have the opportunity to study the importance of amino acid sequence to the structure of the normal hemoglobin protein and normal human red blood cells. You will also investigate the connections between the nucleotide sequence, the physical properties of the hemoglobin polypeptide, the structure of red blood cells, and the physiological effects of a hemoglobin mutation.

References

1. Ingram, V. M. Gene mutations in human hemoglobin: The chemical difference between normal and sickle-cell hemoglobin. *Nature* 180 (1957).

2. Dickerson, R. E., and Geis, I. *Hemoglobin: Structure, Function, Evolution, and Pathology*. Menlo Park, CA: Benjamin/Cummings, 1983.

3. Pauling, L., Itano, H. A., Singer, S. J., and Wells, I. C. Sickle-cell anemia, a molecular disease. *Science* 110 (1949).

4. Klug, W. S., and Cummings, M. R. *Essentials of Genetics*, 2nd ed. Upper Saddle River, NJ: Prentice Hall, 1996.

Introduction

HemoglobinLab will allow you to study the biochemistry of hemoglobin and the relationship of hemoglobin structure and function to the structure and function of human red blood cells. You can be a biochemist who will use techniques such as gel electrophoresis, peptide sequencing, and computer modeling to study hemoglobin structure. Or you can be a molecular biologist and study the relationship between DNA sequence, polypeptide sequence, and hemoglobin structure.

Objectives

The purpose of this laboratory is to:

- Study the structure and function of hemoglobin, the oxygen-transporting protein in human red blood cells.
- Examine the effects of mutations in the globin gene on hemoglobin structure and human red blood cell structure.
- Demonstrate biochemical techniques that can be used to study protein structure.
- Investigate the process of translation by studying the effects of DNA sequence change on the peptide produced using computer modeling.

Before You Begin: Prerequisites

Before beginning HemoglobinLab you should be familiar with the following concepts:

- The relationship between protein structure and function; the levels of protein structure (see Campbell, N. A. Reece, J. B., and Mitchell, L. G., *Biology* 5/e, and Campbell, N. A., and Reece J. B., *Biology* 6/e, chapter 5).
- The use of *TranslationLab* to study transcription and translation.
- How point mutations in a gene can affect protein structure; how a single amino acid substitution mutation in the globin gene results in sickle-cell disease (chapters 5, 14, 17).
- The composition of human blood (chapter 42).
- The use of gel electrophoresis to separate macromolecules (chapter 20).

Assignments

These assignments will enable you to study the effects of mutations in the globin gene for 17 patients and learn how these mutations affect the health of each patient. For your ease in completing each assignment, the background text relevant to the experiment that you will perform is *italicized*, instructions for each assignment are indicated by plain text, and questions or activities that you will be asked to provide answers for are indicated by **bold text**.

The following assignment is designed to help you become familiar with the operation of HemoglobinLab by studying sickle-cell disease.

Assignment 1: Getting to Know HemoglobinLab: Sickle-Cell Disease

Chills, fever, headache, and vomiting are but a few symptoms of the disease called malaria. Malaria is caused by a protozoan, <u>Plasmodium</u> <u>vivax</u>, that lives in tropical countries and reproduces in a mosquito called the <u>Anopheles</u> mosquito. <u>Plasmodium</u> is transmitted to humans when an infected mosquito bites a human and sporozoites, an infectious stage of <u>Plasmodium</u>, enter the human bloodstream and travel to the liver. Sporozoites reproduce in liver cells and release progeny called merozoites into the bloodstream which infect red blood cells, reproduce, and rupture these cells to infect more red blood cells. Individuals who are heterozygous for sickle-cell disease have a higher resistance to malaria than wild type individuals. This resistance occurs because the fragile structure of sickled red blood cells interrupts the life cycle of <u>Plasmodium</u>.

1. Select the Blood Samples view on the input screen for HemoglobinLab. Scroll down the Select Case list and choose patient Miriam Dembele. Read Miriam's case history. Note that her case history is consistent with increased resistance to malaria. Compare Miriam's blood sample with the healthy control sample. **Are there any obvious differences?** Select the Microscope view and make note of any obvious differences in red blood cell structure. **Do any of the red blood cells show phenotypic characteristics of sickle-cell disease? If so, approximately what percentage of her cells show these characteristics?**

2. Select the Gel Electrophoresis view to examine the electrophoretic migration pattern for the β-globin subunits of Miriam's hemoglobin as compared with a control sample from a healthy patient. **Is the migration pattern of Miriam's hemoglobin indicative of a mutation in one of her globin genes? Is Miriam homozygous or heterozygous for this mutation? Explain your answer.**

3. Select the Peptide Sequence view. Click the Find Difference button to identify the amino acid change in Miriam's hemoglobin compared with the normal control hemoglobin. Note: The amino acid in position one is valine and not methionine which is typically the first amino acid for most proteins. This is because the protein is initially translated with a methionine at position one but this amino acid is later cleaved off of the protein. Therefore, the functional protein begins with a valine at position one. Differences in the amino acid sequence of Miriam's hemoglobin protein compared with the normal protein will align at the far left of the screen. **Which amino acid has been substituted for in Miriam's gene? Note the position of this amino acid change. This will be important for identifying the position of the nucleotide change in the globin gene.**

4. Select the Edit DNA Sequence view. Miriam's globin gene sequence appears, compared with the normal, wild-type globin gene sequence. First, you will need to locate the DNA sequence with the triplet ATG that indicates the position of the start codon that would appear on globin mRNA produced by transcription of this gene. You can do this either by scanning the gene by clicking on the double arrows or (more easily) by typing ATG in the Search window and hitting the return key. This will take you to the ATG with the A in nucleotide 87 outlined with a red box. Click on the Bracket Codons button to outline codons beginning at the ATG. Use the single arrow to advance to codon 6 (nucleotides 105–107).

   Click on nucleotide 106—it should now be outlined with a red box—and change the A to a T. Click the Translate button to see a comparison of your custom-mutated protein to Miriam's protein sequence and the normal protein.

   Is this mutation consistent with what you know about the most common mutation that causes sickle-cell disease?

Refer to a codon chart by clicking on the Genetic Code button at the top of the HemoglobinLab homepage and identify the normal codon and the mutated codon that you changed to simulate the amino acid change in Miriam's hemoglobin.

Assignment 2: Gel Electrophoresis of Hemoglobin

The initial diagnosis of a mutation in a hemoglobin gene often involves the interpretation of a patient's clinical symptoms, patient histories, and the results of biochemical tests such as gel electrophoresis and DNA sequencing. The following assignments are designed to illustrate the importance of gel electrophoresis as a technique that can be used to study protein structure.

1. *A number of mutations in the hemoglobin protein result in a mutant protein that demonstrates a faster electrophoretic migration pattern on a gel than the normal protein. Pretend you are a biochemist who is interested in identifying hemoglobin mutations of this kind. You have available to you 17 patients who have donated blood samples to your lab, and your laboratory technician has run gels on the hemoglobin samples from these patients. It is now your job to interpret these gels to identify which patients may contain the mutant forms of hemoglobin that you are interested in learning more about.*

   Select the Gel Electrophoresis view, and examine the electrophoresis pattern for the hemoglobin molecules from each of the patients by clicking on each patient's name.

   Which one of the patients has hemoglobin molecules that show a faster electrophoretic migration pattern than the control molecules?

   Select the Peptide Sequence view and click the Find Difference button to identify the altered amino acid sequence for this patient. What is the mutation that appears? Is this mutation at the N-terminus or C-terminus of the globin polypeptide?

   Provide a possible explanation for why this change in amino acid sequence would cause the mutant protein to show a faster electrophoretic migration pattern than the normal protein.

   Read this patient's case history. This patient's mother has symptoms indicative of a hemoglobin mutation, but the patient's father appears normal.

   Is the gel electrophoresis pattern that you observed for this patient's hemoglobin consistent with his or her family history? Explain your answer.

   After you meet with this patient and discuss your interest in his or her blood, the patient tells you that his or her mother was born and raised in Hiroshima, Japan.

Because you are a well-rounded person with a strong knowledge of world history, what might you consider to be the cause of this patient's hemoglobin mutation?

2. *Certain mutations in the β-globin gene result in altered amino acid sequences in the hemoglobin molecule that produce a protein with an increased affinity for binding to oxygen. One example of such a mutation produces a molecule called hemoglobin Yakima. Yakima involves an amino acid substitution mutation at position 99, where aspartic acid is replaced by histidine. Individuals with these mutant forms of hemoglobin often show redder-than-average complexions. Select the Blood Samples view and scroll through the patient case histories searching for the patient whose complexion matches this description.*

   Once you think you have found this patient, select the Microscope view and evaluate the patient's red blood cells for any obvious defects. Select the Gel Electrophoresis view.

   Does the migration pattern of this patient's hemoglobin indicate a mutation in the protein? If your answer is yes, does this patient appear to be homozygous or heterozygous for this mutation?

   Select the Peptide Sequence view and click the Find Difference button to identify the altered amino acid sequence for this patient.

   What is the mutation that appears? Is this mutation indicative of hemoglobin Yakima? Provide reasons why you think this mutation may increase the affinity of oxygen for binding to hemoglobin Yakima.

Assignment 3: Peptide Sequence Analysis of Hemoglobin

As powerful as gel electrophoresis is as a technique, changes in the amino acid sequence of a protein can be definitively determined only by analyzing DNA and peptide sequences. The following assignments will help you understand how these techniques can be applied to study protein structure.

1. **Select the Blood Samples view and click on patient Rhonda Emolina. Compare the color of Rhonda's blood with that of the healthy control blood. Is the color of Rhonda's blood consistent with the conditions described in her patient history?**

   To determine the cause of Rhonda's anemia (abnormally low number of red blood cells), select the Gel Electrophoresis view.

   Does Rhonda's hemoglobin migrate differently than the healthy control sample?

Because the results of this gel electrophoresis experiment are inconclusive in determining whether the cause of her anemia is due to a hemoglobin mutation or another problem such as an iron deficiency, more information about the sequence of Rhonda's hemoglobin protein needs to be considered.

Select the Peptide Sequence view and click on the Find Difference button to determine whether Rhonda may contain a mutated version of hemoglobin. Is there a mutation? If so at which position and what amino acid is changed?

2. *Many invariant amino acid positions around the heme group have been identified in vertebrate and invertebrate hemoglobins. Changes in these invariant amino acids, many of which form the hydrophobic pocket around the heme group, result in a number of very serious blood diseases including a wide variety of anemias. Anemia is a condition that involves an abnormally low oxygen-carrying capability of the blood. Anemias can occur if the number of red blood cells in an individual is low. Anemias are also caused by the production of abnormal hemoglobin molecules (such as sickle-cell anemia, and thalassemia) and inadequate hemoglobin content in red blood cells. Common symptoms of anemias are decreased blood oxygen content, fatigue, shortness of breath, pale skin, and cool body temperature.*

One of the invariant amino acids in β-globin is mutated in the form of hemoglobin called hemoglobin Hammersmith. This invariant amino acid is located at position 42 in the β-globin polypeptide. This mutation results in an unstable hemoglobin that cannot hold and position the heme group in the proper orientation for oxygen binding.

Select the Peptide Sequence view and compare the hemoglobin sequence for each female patient with the control sequence by clicking on the Find Difference button until you identify the female patient with hemoglobin Hammersmith. (Be sure to begin your search by starting at the first amino acid in the protein).

Which amino acid is substituted for in this patient?

Use the Edit DNA Sequence view to identify the codon for this amino acid. Alter this codon by changing positions in the codon until you have recreated the Hammersmith mutation. Refer to the codon chart in HemoglobinLab to confirm the mutation that you created.

Assignment 4: Group Assignment

In the previous exercises we have considered alterations in hemoglobin structure created by base-pair substitution mutations in the globin gene. Working together in a group of four or five students complete the following assignment which considers other types of mutations.

1. Select patient Juan Rodriquez. Click on the Blood Samples view and read his history. Note any differences in the color of Juan's blood compared with the control cells. Select the Microscope view and note the appearance of Juan's red blood cells compared with the control cells. Select the Gel Electrophoresis view and describe the electrophoretic pattern of Juan's hemoglobin.

   Based on this electrophoretic pattern, develop at least two hypotheses that could explain this observation.

2. To determine if any of your hypotheses are correct, select the Peptide Sequence view. Click on the Find Differences button.

   Compare the amino acid sequence of Juan's hemoglobin with the control sequence.

   a. Once you have determined the first amino acid difference, refer to the codon chart in HemoglobinLab and identify the codon for the amino acid on the normal protein sequence. Select the Edit DNA Sequence view to modify this codon and recreate this mutation.

   b. Return to the Peptide Sequence view. Use the double arrows to examine the rest of Juan's hemoglobin. Record all differences in amino acid sequence that you observe.

      Do the results of this examination confirm or refute your hypothesis? If necessary, formulate a new hypothesis to account for this observation.

   c. **Return to the Edit DNA Sequence view. Based on your hypothesis, alter the codon sequence of Juan's globin gene until you have identified the nucleotide change(s) that have occurred in Juan's globin gene.**

   d. **What is an aplastic crisis? What are common causes of aplastic anemia? Consider Juan's patient history. How might his aplastic crisis have resulted from the hemoglobin mutation that he has?** If necessary, refer to an anatomy and physiology textbook in your library to answer this question.

EvolutionLab

Background

Evolution is the study of how modern organisms have descended from the earliest life-forms and of the genetic, structural, and functional modifications of a population that occur from generation to generation. The ability of a population of organisms to respond to change in their environment and survive and reproduce by developing the characteristics or modifications necessary for survival is known as **adaptation**. Understanding how life evolves is a central concept in biology. The incredible diversity of living organisms that exists and their adaptations to their environment are the direct result of evolution that has occurred over very long periods of time.

Many scientists have contributed to our modern-day understanding of evolution. One of the early pioneers of evolution was the French biologist Jean Baptiste de Lamarck. Working in the early 1800s, Lamarck compared fossilized forms of invertebrate organisms and noted that younger fossils showed advanced structural characteristics compared with older fossils. Lamarck suggested that younger fossils displayed adaptations, or modifications of structures, that were indicative of an increase in an organism's complexity. He proposed that these modifications arose from the use or disuse of body parts. He also hypothesized that certain structural features developed in response to an organism's environment and that many of these acquired characteristics could be inherited by offspring.

One classic example of such modifications suggested by Lamarck involves the long neck of giraffes. He proposed that the earliest giraffes were relatively short-necked animals that were forced to stretch their necks to eat their desired food, leaves that grew at the tops of trees. Over time, such stretching would produce animals with small increases in the length of their necks. This trait would be passed on to future offspring, which would also show increased neck length as they searched for food. Over many generations, these adaptive changes would produce giraffes with a greatly increased neck length compared with their early ancestors. At the time, however, a biological basis for Lamarck's hypotheses remained unclear.

The English naturalist Charles Darwin is generally recognized as the father of evolution. Many of Darwin's now-famous hypotheses were developed as a result of his 1831 voyage on HMS *Beagle*, a ship commissioned by the British government. In this historic voyage, the *Beagle* traveled to many coastal areas of South America. The location that resulted in Darwin's most important observations was the Galápagos Islands in the South Pacific Ocean, off the northwestern coast of South America. Darwin studied populations of Galapagos tortoises and observed their feeding habits regarding the fruit of prickly pear cactuses. Darwin noticed differences in tortoise variety and prickly pear cactus growth that appeared on the various islands of the Galápagos. For example, he observed that on islands without tortoises, prickly pear cactuses grew fruit very close to and, in some cases, on top of the ground. But on islands with tortoises, the prickly pear cactuses showed tall trunks that elevated the fruits of the cactus out of the reach of ground-dwelling tortoises. These observations led Darwin to consider interrelationships between tortoises and their food source. He also observed that many of the different islands of the Galapagos contained distinctly different varieties of finches. It was from Darwin's observations of finches that many of his pioneering theories of evolution by natural selection were developed. Most of these theories continue to provide the basis for our current-day understanding and study of evolution.

Several years after Darwin's initial voyage to the Galápagos Islands, he developed several hypotheses that he felt explained how diverse species of tortoises and finches may have developed on the different islands. He realized that although the populations of these organisms could grow rapidly, tortoise and finch populations did not increase in an uncontrolled manner; rather, the populations of these animals were somehow maintained in a relatively constant size. Darwin noted that certain individual organisms in a population showed subtle structural and functional differences. Because the habitat on each island appeared to stay the same without dramatic changes in environmental conditions such as food supply, he reasoned that there would be competition among individuals for these conditions. In response to this competition, Darwin suggested, some individuals in a population were likely to develop certain structural and functional characteristics, or traits, as a way to increase their odds of survival. Developing the traits that allow individuals to survive and reproduce in an environment is called **adaptation**. Assuming that certain structural differences were essential for survival, he reasoned that these organisms would be superior at surviving in an environment where a population was competing for limited resources. Such favorable structural differences would be maintained or inherited over many generations, while less favorable structural differences were likely to be lost. This line of thinking is often referred to as survival of the fittest because, over time, a population would consist of organisms that inherited those favorable traits that are best suited for survival of the population.

Darwin proposed that evolution of a population in this fashion occurred by **natural selection**, because the environment selects for individuals with traits that allow them to adapt to a given environment. Because of the greater **fitness** of these organisms, they would be more likely to produce the greatest number of offspring, whereas those organisms that possessed less favorable abilities would be more likely to die. The term fitness is used to describe the reproductive success of an organism. Environmental factors that influence adaptation can be nonliving (**abiotic**) environmental components or living (**biotic**) components, such as competition of organisms for a food source or mating behaviors of a population. Over many generations, natural selection results in changes in the individuals of a population that allow the population to evolve. It is important to realize that natural selection happens to individuals, but that evolution is a function of the changes of the collective individuals in a population over time and not a set of changes that are observed in a given individual.

Darwin's pioneering observations and hypotheses were chronicled in his famous book, *On the Origin of Species by Means of Natural Selection*, which was originally published in 1859. Around the same time that Darwin was developing his important theories, Alfred Russel Wallace was also independently formulating lines of reason that supported evolution by natural selection. At the time of Wallace's and Darwin's work, and for many years afterward, the significance of evolution by natural selection was not fully realized by biologists. In more recent times, however, the concept of evolution by natural selection has been substantiated and validated by the evidence provided from fossil records, ancestry studies of animals that consider embryological stages of development in related organisms, comparative anatomy, and most recently, DNA sequence comparison of related organisms.

The biological basis for the inheritance of traits and the principles of evolution established by Lamarck, Darwin, Wallace, and others was significantly advanced by the work of the Augustinian monk Gregor Mendel. Mendel studied inheritance of traits using the garden pea, *Pisum sativum*. In the 1860s, remarkably, without an understanding of chromosomes and genes, Mendel established that units of inheritance existed to transmit information on traits from parents to offspring. Many years later, the **gene** was identified as the unit of inheritance. Mendel also established the fundamental rules and patterns by which traits are inherited that continue to form the basic principles of genetics that are followed in modern-day genetics laboratories. Mendel's discoveries, and subsequent work on gene structure and function, established an explanation for how organisms can change over time to produce individuals with desirable heritable features that increase the odds for population survival. Mendel recognized that organisms typically contain alternate forms of genes, called **alleles**. Within a given population, several different alleles for a given trait may exist (for example, eye color as a trait is determined by combinations of different alleles for the protein melanin). One way that alleles arise is through mutations—changes in the nucleotide sequence of a gene. Mutations are largely responsible for the production of new alleles and the tremendous genetic diversity that exists in virtually all populations of organisms. The sum total of all genes in a population is defined as a population's **gene pool**. The spread of new alleles within a population and between different populations forms the molecular basis for evolution by natural selection.

Changes in the **genotype** of an organism as mutations arise create changes in **phenotype**—the behaviors, structural appearances, and physiology of an organism. Thus, natural selection produces a range of different genotypes and phenotypes in a population over time. Population genetics involves studying the frequency and distribution of gene inheritance within a population and the genetic variations that arise within populations. Evolution by natural selection results from generational changes in population frequencies of alleles that influence the phenotypic characteristics of a population. Allele frequency is a measure of the abundance of different alleles in a population. Evolution occurs over many generations as individuals with different alleles reproduce. As individuals in a population breed with each other and with individuals from other populations, allele frequency will change as each population gains some alleles and loses others. This concept is known as **gene flow**. Hence, evolution results from inevitable changes in large numbers of genes within a population over many generations. There are three modes by which the heritability of a trait in a population can be influenced by natural selection: **directional selection**, **stabilizing selection**, and **diversifying selection**. These modes are discussed in the Assignments sections of EvolutionLab.

Evolution is a central theme that unifies and connects virtually all disciplines of biology. However, because evolution occurs over long spans of time and is difficult to observe directly, the mechanisms that drive natural selection do not lend themselves to experimentation during a few hours or even a semester spent in the biology lab. In EvolutionLab, you will have the opportunity to simulate experiments in evolution to help you understand how traits of an organism can change in response to different biological and environmental conditions. You will study how beak size and population numbers for two hypothetical populations of finches on two different islands can evolve in response to factors that you will manipulate by changing environmental conditions on these islands. EvolutionLab is not an exact simulation of finch beak evolution; rather it is a model designed to demonstrate important principles of adaptation by natural selection. Later you will have the opportunity to study Mendelian genetics in the fruit fly, *Drosophila melanogaster*, using FlyLab. You will also have the opportunity to use PedigreeLab to study how genetic traits are inherited across several generations of humans.

References

1. Darwin, C. *On the Origin of Species by Means of Natural Selection, or the Preservation of Favored Races in the Struggle for Life.* New York: New American Library, 1963.

2. Gould, S. J. "The Evolution of Life on the Earth." *Scientific American*, October 1994.

3. Grant, P. R. "Natural Selection and Darwin's Finches." *Scientific American*, January 1991.

4. Weiner, J. *The Beak of the Finch: A Story of Evolution in Our Time.* New York: A. Knopf, 1994.

Introduction

EvolutionLab will allow you to study important principles of evolution by examining small populations of finches on two different islands, "Darwin Island" and "Wallace Island." You will manipulate important parameters that influence natural selection and then follow how your changes influence the evolution of beak size and population numbers for the two different populations of finches over selected time intervals. One obvious advantage of this simulation is that you can quickly and easily create new parameters and model evolutionary changes over long time periods.

Objectives

The purpose of this laboratory is to:
*   Help you develop an understanding of important factors that affect evolution of a species.
*   Demonstrate important biological and environmental selection factors that influence evolution by natural selection.
*   Simulate how changes in beak size and other characteristics of finch populations influence evolution of beak size and population numbers.

Before You Begin: Prerequisites

Before beginning EvolutionLab you should be familiar with the following concepts:

- Evolution and the importance of natural selection as a central concept of Darwinism (see Campbell, N. A., Reece, J. B., and Mitchell, L. G., *Biology* 5/e, and Campbell, N. A., and Reece J. B., *Biology* 6/e, chapter 22).
- Biological and environmental factors that influence natural selection (chapter 23).
- Evolutionary principles associated with population genetics (chapter 23).

Assignments

The finches on Darwin and Wallace Islands feed on seeds produced by plants growing on these islands. There are three categories of seeds: soft seeds, produced by plants that do well under wet conditions; seeds that are intermediate in hardness, produced by plants that do best under moderate precipitation; and hard seeds, produced by plants that dominate in drought conditions. EvolutionLab is based on a model for the evolution of quantitative traits—characteristics of an individual that are controlled by large numbers of genes. These traits are studied by looking at the statistical distribution of the trait in populations and investigating how the distribution changes from one generation to the next. For the finches in EvolutionLab, the depth of the beak is the quantitative trait. You will investigate how this trait changes under different biological and environmental conditions.

You can manipulate various biological parameters (initial beak size, heritability of beak size, variation in beak size, clutch size, and population size) and two environmental parameters (precipitation, and island size) of the system, then observe changes in the distributions of beak size and population numbers over time.

For your ease in completing each assignment, the background text relevant to the experiment that you will perform is *italicized*, instructions for each assignment are indicated by plain text, and questions or activities that you will be asked to provide answers for are indicated by **bold text**.

Assignment 1: Getting to Know EvolutionLab: The Influence of Precipitation on Beak Size and Population Numbers

1. *The first screen that will appear in EvolutionLab presents an initial summary (Input Summary) of the default values for each of the parameters that you can manipulate.*

   Notice that default values on both islands are the same. Click on the Change Inputs button at the left of screen to begin an experiment. A view of initial beak size will now appear.

   In the Change Input view you can change the biological and environmental parameters in EvolutionLab to design an experiment. This first experiment is designed to study the influence of beak size on finch population numbers. For finches, deep beaks are strong beaks, ideally suited for cracking hard seeds, and shallow beaks are better suited for cracking soft seeds.

Develop a hypothesis to predict how changes in beak size will affect population numbers for these finches. Test your hypothesis as follows:

a. Begin by setting the initial beak size on the two islands to opposite extremes. Leave the initial beak size on Darwin Island at 12 mm and click and drag on the slider to change the initial beak size on Wallace Island to 28 mm. Note the change in beak size that appears on the graphic of each finch. Click the Done button to return to the input summary view. Notice the new input value (28 mm) for beak size on Wallace Island while beak size on Darwin Island remains at the default value of 12 mm. Click on the Precipitation button to view the distribution of seed types on both islands. Use the popup menu in the lower left corner to select a value of 300 years, and run the simulation by clicking the Run Experiment button.

b. Once the experiment has run, you will be in the Beak Size view.

Look at the plots of average beak size over time. What do you observe? Do you notice any trends in beak size? Click on the Population button and look at the plots of population numbers over time. What changes do you see? Do the two islands differ? Does the data support or refute your hypothesis?

Data from the Beak Size view, and Population view are shown in tabular form in the Field Notes view. Click on the Field Notes view. A table showing each year of the experiment, mean beak size, and population will appear.

Click on the Histograms button. These are plots of surviving birds and total birds plotted against beak size. Click and drag the slider to advance the years of the plot and to see how beak size on each island may have changed over time. Note how the distributions of beak size change over time.

What happened to beak size on Darwin Island compared to Wallace Island over time? Is this what you expected? Why or why not?

Click on the Input Summary button to see a table of your input values for this experiment. Data from each of the views that you just looked at can be saved in a virtual notebook by clicking on the Export Data button. A new window will appear. You can type comments on your results in this window. To print your lab notes, click on the Export Notes button. A new window will appear. You can now save your notes to disk and/or print a copy of your lab notes using the print feature of your browser software.

2. *This experiment is designed to explore the effect of precipitation on finch beak size and population numbers.*

Click the New Experiment button, click the Change Inputs buttons then click the Precipitation button. Recall the relationship between precipitation and seed growth. There are three categories of seeds: soft seeds, produced by plants that do well under wet conditions; seeds that are intermediate in hardness, produced by plants that do best under moderate precipitation; and hard seeds, produced by plants that dominate in drought conditions.

Develop a hypothesis to consider how a decrease in precipitation on Darwin Island might affect beak size and develop a hypothesis to explain how a decrease in precipitation might influence population numbers for these finches over time. Test your hypotheses as follows:

a. Leaving all other parameters at their default values, decrease precipitation on Darwin Island to 0 mm. Notice how the distribution of seeds produced on Darwin Island changes as you change precipitation. Set the experiment to run for 100 years, then run the experiment. Compare beak size and population numbers for the finches on Darwin Island over 100 years. Scroll down the Field Notes view to observe the data recorded over 100 years. Use the Beak Size and Population buttons to view the effect of your experiment on each of these parameters.

Did you notice any trends in the distributions of beak size? What did you observe? Did you notice any trends in population number? What did you observe? Explain your answers.

Run another experiment for 200 years by clicking on the Revise Experiment button. Use the popup menu at the lower left corner of the screen to select a value of 200 years, then click the Run Experiment button. Repeat this experiment for 300 years.

What changes did you observe in beak size and population numbers? Do these results confirm or refute your hypothesis? If necessary, reformulate your hypothesis and test this hypothesis.

b. Perform the same experiment for both Wallace Island and Darwin Island simultaneously.

Did you notice any differences between precipitation, changes in beak size, and population numbers for the finches on Wallace Island compared with those on Darwin Island? Explain your answers.

c. **Develop a hypothesis to consider how an increase in precipitation on Darwin Island might influence the evolution of beak size.**

Click the New Experiment button, return to the Change Inputs view then increase the precipitation on Darwin Island fourfold while leaving precipitation on Wallace Island at the default value. Run this experiment for 300 years to test your hypothesis.

What did you observe? After you have observed the data for this experiment, rerun this experiment. Look at the output results in the Beak Size and Population views. Do you notice any differences in this rerun compared with the previous run? Are the general trends observed in this run the same as the previous run? Explain your answers. Run and rerun each experiment for 100, 200, and 300 years. Perform another experiment to test your hypothesis by increasing precipitation on Wallace Island to 50 cm/year and increasing beak size to 28 mm. Run an experiment for 300 years and describe your results. Do these results support your hypothesis?

d. Decrease beak size on both of the islands to an intermediate value. Decrease rainfall on one island to a value close to zero. On the other island, increase rainfall close to the maximum value. Run the experiment for 300 years.

Were the effects on each island the same or different? What did you observe? Were these the results you expected? Explain your answers to justify what is happening to finches on each island. Be sure to provide explanations for any differences in beak size and population numbers that you observed when comparing finches on both islands.

Assignment 2: Modes of Natural Selection

*There are three primary ways by which natural selection can influence the variation of a trait in a population. The three modes of selection are **directional**, **stabilizing**, and **diversifying**. In directional selection, changing environmental conditions can favor individuals with phenotypes that are at opposite extremes of the distribution range for a given trait (for example, finches with very deep beaks or finches with very shallow beaks). Stabilizing selection, unlike directional selection, selects against individuals with extreme phenotypes and favors individuals with more intermediate or average values for a given trait. Disruptive selection favors individuals at both extremes of the distribution range for a trait while selecting against individuals with average values for the trait. The following assignment is designed to help you understand how environmental changes can result in different modes of natural selection.*

1. Leaving all of the other settings at their default values, change the rainfall on Wallace Island to 50 cm/year and the rainfall on Darwin Island to the minimum possible value (0 cm/year). Run the simulation for 300 years.

    a. **Look at the plot of beak size over time. What type of selection is taking place on Wallace Island? On Darwin Island? Explain your answer.**

    b. **Using the numbers from the field notes, calculate the average $\underline{R}$ (the difference in mean beak size from one generation to the next) in the**

86

first ten years of the simulation and the last ten years for both populations.

c. Look at the plot of finch population over time. Explain the reason for any differences in population numbers between the two islands.

Assignment 3: Effect of Clutch Size

Clutch size is the number of eggs that a female bird lays in her nest. In the EvolutionLab simulation model, birds mate for life and live for one year, and each female produces only one clutch of eggs per year. For a mating pair to replace themselves, they must produce at least two offspring; this is why clutch size is set to a minimum of two eggs. The maximum clutch size of thirty eggs is unrealistic; however, most bird species produce more than one clutch in their lifetime, so a total of thirty or more offspring per mating pair is possible. You can use the sliders to change the mean clutch size for each island population. As you move the slider, the number of eggs in the nest will change to reflect the value that you have chosen.

1. Set the clutch size to 6, the rainfall to 37 cm, and the initial beak size to 25 mm on both islands (keep everything else at the default values). Run this simulation for 300 years and repeat the run two or three times.

   Did you notice anything odd? (If not, try again until you do). Propose a hypothesis to explain this result. Leaving the rainfall and beak sizes alone, what parameters would you change to prevent this? What parameter would you change to increase the likelihood of this happening? What type of selection is happening during the first several years of this experiment? If these birds were capable of assortative mating, what might happen on one island?

Assignment 4: Effect of Island Size

*The size of the living area for any population can strongly influence population numbers for organisms that live within that environment. The maximum number of organisms from a given population that an environment can support is known as the **carrying capacity** of that environment. Island size is one factor that can determine the carrying capacity of finches on each island. For the purposes of this simulation, the islands are assumed to be roughly circular and island size is represented as the radius of the island in kilometers. The size of each island remains constant throughout the simulation unless you choose to change this parameter. Although changing the entire size of an island is not something that could easily be done in real life, habitat changes and reducing the living environment for a population are real changes that occur through processes such as land development, and pollution. The following assignment is designed to help you learn about the influence of island size on the carrying capacity of finches on Darwin and Wallace Islands.*

1. **Develop a testable hypothesis to predict what effect an increase in island size will have on beak size and finch populations.**

Test your hypothesis as follows: Begin your experiment by leaving all other parameters at their default values. Select the Island Size input and use the sliders to increase the size of either Darwin or Wallace Island. As you move the slider, the island image will change to reflect the values you have chosen.

What effect did this change in island size have on finch population? What effect did this change in island size have on beak size? Are the results what you expected? Explain your answers. Perform a new experiment to learn about the effects of a decrease in island size on beak size and finch populations.

2. Based on the previous experiments, consider possible parameters that you could manipulate which would prevent the changes in population size and beak size that you observed from occurring.

   Test the effect of these parameters to influence population size and beak size by designing and running experiments to confirm or refute your answers.

Assignment 5: Variance

Variance is a measure of how different the phenotype is from one bird to the next. If the variance of a trait is large, then there will be large differences in phenotypes among birds for a given trait. If, on the other hand, variance is low (near zero), then all of the birds will be very similar to one another. This parameter determines two values for the simulation. First, it determines the initial population variance—variance in beak size for the entire population. The population variance may change during the course of the simulation. The variance parameter also determines the sibling variance—variance in beak size among individuals with the same parents. The sibling variance is a measure of how much variability is inherent in a trait. If you look at the plots of offspring beak size versus the midparent value for the heritability parameter, the sibling variance represents the amount of dispersion of the points around the regression line. In EvolutionLab, the sibling variance remains constant throughout the simulation. You can use the sliders to change the variance for each island population. As you move the slider, the width of the bell-shaped probability distribution will change to reflect the value that you have chosen.

1. **For the finches on one of the islands, develop a hypothesis to consider the effects of changes in variance on population numbers.**

   Begin by decreasing variance to a value close to zero. Run your experiment for 300 years.

   Explain the results of this experiment. Perform a similar experiment for an increase in variance. Describe your results. Can you reverse any of the trends in population number by changing other variables such as precipitation?

2. Try another experiment in which you increase variance on Darwin Island to a value close to 2.0. Decrease variance on Wallace Island to a value close to 0.1. Equally change precipitation on both islands to a value close to zero. Run the experiment for 300 years.

   What were the results of this experiment? Explain why the differences in variance on each island produced these results. What do these results tell you about the benefit of a population with a broad variance compared to a population with narrow variance? Which condition do you think is most desirable for maintaining a diverse population and for minimizing a population's risk of extinction?

Assignment 6: Extinction

Now that you have manipulated many of the biological and environmental factors in EvolutionLab, consider how these factors could lead to extinction of the finch population on either island.

1. **What conditions would lead to extinction of a finch population? Which of the parameters is most important in determining whether a population goes extinct? Can you describe at least three different sets of conditions that will lead to extinction?**

2. **Design and perform experiments that will confirm or refute your answers. Compare your answers with those of other classmates. How do they compare? If your classmates have different conditions that lead to extinction, design and perform an experiment to validate the ability of those conditions to cause extinction.**

Assignment 7: Group Assignment: Influence of Heritability

Heritability is a measure of the genetic contribution to a phenotype of an organism. If the heritability of a trait is large (near one), then the progeny of a mating will be more similar to their parents than unrelated individuals. If heritability is low (near zero), then the environment will be more important in determining the phenotype and the progeny will vary around the average value for the population. One method for measuring heritability is to plot the beak depth of each bird versus the average beak depth of each bird's parents (midparent value). The slope of the line produced by this plot represents heritability.

You can use the sliders to vary the heritability of beak size on each island population. Heritability remains constant at this value throughout the simulation. As you move the slider, the graph of offspring beak size versus the midparent value will change to illustrate the heritability value that you are selecting. The mean value for the plot is equal to the initial mean beak size that you have chosen. The amount of dispersion of the points around the plot is determined by your selection for variance in beak size.

Work together in a group of four students to perform the following assignment, which is designed to help you understand the influence of heritability on population numbers. Click on the Heritability button and read the description of heritability and heritability plots.

1. **What effect do you think heritability will have on the evolution of finch beak size? As a group, discuss the importance of heritability. Design at least one hypothesis to explain how changes in heritability might affect the evolution of beak size and then use EvolutionLab to test your hypotheses. What did you do, and what were the results?**

    a. **Looking back over the experiments you did with heritability and variability, would you say that changes in these properties of the populations cause similar effects or different effects on evolution of the finches' beaks? Does either of these parameters affect the direction of evolution? The end point of evolution? The rate of evolution? Discuss your answers with students in your group.**

    b. **If the heritability of one population is set to 0.75 while the other is set to 0.25, could the rate of evolution on the two islands still be the same if there were differences in variability between the two islands? Try it and see if you can find values for variability that compensate for the differences in heritability. What is going on here and what is the connection between heritability and variability?**

PopGenLab

Background

For some individuals, evolution is a controversial topic. For most biologists, evolution is a central, unifying theme that incorporates many fields of biology including genetics, molecular biology, anatomy, and ecology among other disciplines. You may have already used EvolutionLab to study principles of evolution by natural selection in simulated populations of finches. Even if you have not studied evolution in detail, you are likely to know that changes in the genetic material of an individual in a population can and do occur through mutations. But remember that individual organisms do not live long enough to undergo significant genetic change whereby biologists consider an individual to have evolved. Genetic changes that occur in generations of populations over long periods of time are the basis for evolution of a species.

In a population of any species, there are typically individuals that show differences in phenotype for a particular trait. These differences represent genetic variation in the population. For example, the finches Charles Darwin studied on the Galápagos Islands showed variations in beak size. Of course, the underlying basis for phenotypic differences is the **genotype**, or genetic composition of an organism. **Mutations**, both random and induced, are the sources of new genes and new alleles that cause the heritable variation that is essential for evolution of a species and its population. The total range of genetic variation, due to **alleles** for all genes in a population, is known as the **gene pool** for that population. Geneticists interested in population genetics study allele frequencies as a way to predict whether a population is evolving. When there are multiple alleles for a particular gene, it is sometimes possible to determine the relative proportion of each allele in a gene pool. If all individuals in a population are homozygous for an allele being studied, that allele is known as a fixed allele.

What is the relationship between the frequency of alleles and the frequency of genotypes in a population? In 1908, two geneticists, G. H. Hardy and W. Weinberg, independently proposed an equation to relate allele frequency to genotype frequency. If one is studying two alleles called A and a for a particular gene at a known locus, the genotype frequencies of the possible combinations of these alleles—AA, Aa, and aa—can be determined from the equation $p^2 + 2pq + q^2 = 1.0$, where p can be designated as the frequency of the A allele, and q can be designated as the frequency of the a allele. Consider a population of 100 individuals, each of which has two alleles for this gene. The sum of p and q represents 100% of the alleles for this gene in the population. Therefore, if 80 percent of the alleles are A, then p would be 0.8. Twenty percent of the alleles would then be a; therefore q would be 0.2. The genotype frequencies would be $p^2 = (0.8)^2 = 0.64$ for the frequency of AA homozygotes, $2pq = 2(0.8)(0.2) = 0.32$ for the frequency of Aa heterozygotes, and $q^2 = (0.2)^2 = 0.04$ for the frequency of aa homozygotes. We would predict 64 AA, 32 Aa, and 4 aa genotypes in this population of 100 individuals.

How does an understanding of the **Hardy-Weinberg theorem** help us understand population genetics? An understanding of allele frequency in a population can tell us whether evolution is occurring in a population or if that population is in a state of genetic equilibrium. If a population is not evolving, then allele frequency and genotype frequency will remain the same from generation to generation and the genotype frequencies will correspond to the Hardy-Weinberg prediction. The allele and genotype frequencies of a parental generation will match those of their offspring. Such nonevolving populations are said to be in a state of Hardy-Weinberg equilibrium. However, to truly understand population genetics, we need to consider the factors that influence allele and genotype frequencies. Hardy and Weinberg proposed that there are a number of conditions that must be held constant in a population if allele frequencies and genotype frequencies are to remain constant over several generations. These conditions are as follows:

(1) Large breeding population. The effect of random changes in allele frequencies (**genetic drift**) is greatly reduced in a population with a large number of individuals.

(2) Random mating. Individuals in a population show no preference for mating with other individuals of a particular phenotype.

(3) The alleles under study do not mutate. Alleles are not being mutated to create new alleles, which would change the gene pool and alter genotype frequencies in a population.

(4) There is no migration of individuals into or out of the population; therefore, the gene pool will not change due to migration.

(5) There is no selective advantage for any individual. If all individuals in a population have an equal chance of surviving and reproducing, then all of the genotypes represented in the population are equally viable and all alleles should be inherited equally.

These five conditions are required to maintain Hardy-Weinberg equilibrium. If all conditions are met in a population, then no change in allele or genotype frequency will occur in that population. Therefore, we can use these criteria to determine if changes in a gene pool are occurring in a population—a process known as **microevolution**. It is important to understand how changes in Hardy-Weinberg criteria can result in microevolution. Many of these criteria will be further described in the assignments that accompany this lab as you design experiments that will help you learn how these criteria can influence population genetics.

It is possible to predict the proportions of individuals in different generations that will show certain phenotypes when one is studying very large populations and the population size is effectively infinite. It can be difficult to accurately study population genetics in small populations. Small populations may arise from larger populations due to events such as a natural disaster—for example, loss of habitat in an extreme weather condition such as a fire may isolate populations of organisms that were once part of a larger population. However, in a small population, the proportions of individuals with a certain phenotype can be strongly influenced by random events in the gene pool. These random events are called **genetic drift**. Genetic drift can make it very difficult to accurately predict phenotypic frequencies in a small population. For example, random variation in survival and number of offspring among genotypes could, over time, change the proportions of genotypes in the population. The effects of chance events such as these are greatly minimized in large populations.

One particularly well-characterized example of natural selection in a population involves British moths called peppered moths (*Biston betularia*). These moths are found in two different colors or morphs. One morph has light, almost white-colored wings with small flecks of brown, while the other morph is predominantly black and brown in color. Because birds are predators of peppered moths, wing color is an important camouflage for moths. White morphs can blend well into the light, peppered appearance of lichen-colored trees.

In the early 1800s the black morph began to appear in greater frequencies in cities throughout England, particularly highly industrialized cities, while the white morph dominated populations in rural areas. The increased frequency of the black morph coincided with the industrial revolution occurring in England. It was determined that the increase in soot released from coal-burning plants killed many species of lichen growing on trees in industrialized areas of England and blackened tree bark to a much darker color than lichen-covered trees in rural areas. As a result of this pollution-induced change in the environment, white morphs became much more visible and more vulnerable to predation by birds. Black morphs began to dominate populations in industrialized areas because they were less likely to be seen and eaten by birds. In some industrialized areas, genotype frequencies for black morphs approximated 100% of the population. The selective advantage by black morphs disrupted Hardy-Weinberg equilibrium conditions for moths in industrialized areas. The story of *Biston betularia* is a classic case of natural selection in a population and an example of how evolution can proceed via strong selection pressures.

You will use PopGenLab to learn how changes in important parameters of population genetics can influence evolution in simulated populations of moths that resemble peppered moths. You are provided with moths living in tree stands. A single gene with two alleles controls wing color of these moths, and each genotype produces a different color pattern. Moth survival depends on the insect's ability to blend against the bark of the trees they are living in. Different tree types are provided with bark colors that match the colors of the moths in the simulation. The experiments that you set up and analyze in PopGenLab will provide you with an important understanding of the factors influencing Hardy-Weinberg equilibrium and natural selection. By varying parameters such as allele frequencies and survival rates of each genotype, population numbers, population carrying capacity, mating patterns, and the frequency of population crashes due to natural disasters, you will design experiments to help you understand how each parameter can affect evolution within the population of moths.

References
1. Atherly, A. G., Girton, J. R., and McDonald, J. F. *The Science of Genetics*, 1st ed. New York: Saunders, 1999.

2. Freeman, S., and Herron, J. C. *Evolutionary Analysis*, 2nd ed. Upper Saddle River, NJ: Prentice-Hall, 2000.

3. Smith, J. M. *Evolutionary Genetics*, 2nd ed. New York: Oxford University Press, 1999.

4. Stearns, S. C., and Hoekstra, R. F. *Evolution: An Introduction*, 1st ed. New York: Oxford University Press, 2000.

Introduction

In this laboratory, you will perform simulations of experiments designed to study factors that can lead to changes in genotype frequency in a population resulting in population genetic changes that influence the evolution of moths. By manipulating genetic and environmental parameters that influence the genetics of these moths, you will learn about important principles of population genetics and the factors that can affect population genetics for any species.

Objectives

The purpose of this laboratory is to:
- Demonstrate important relationships between the many factors that influence population genetics in a simulated population of moths.
- Simulate the effects of changing conditions such as genetic drift, mating patterns, mutations, migration, and genetic bottlenecks, which can lead to microevolution in a population.
- Investigate how changes in genetic and environmental parameters that affect population genetics can affect the evolution of a population.

Before You Begin: Prerequisites

Before beginning PopGenLab, you should be familiar with the following concepts:
- Basic terminology, principles, and genetic crosses based on Mendelian genetics (see, N. A., Reece, J. B., and Mitchell, L. G. *Biology*, 5/e, and Campbell, N. A., and Reece J. B., *Biology* 6/e, chapter 14).
- Evolution of a species by natural selection, and the effect of changes in allele and genotype frequencies on evolution in a population (chapter 23).
- Important conditions that influence microevolution including genetic drift, gene flow, mutation, modes of selection in a population, genetic bottlenecks, and random and nonrandom mating (chapter 23).
- Modes of selection in a population (chapter 23).
- Applying the Hardy-Weinberg principle (theorem) to calculate genotype frequency for a population (chapter 23).

Assignments

For your ease in completing the following assignments, the background text relevant to the experiment that you will perform is *italicized*, instructions for each assignment are indicated by plain text, and questions or activities that you will be asked to provide answers for are indicated by **bold text.**

The following assignment is designed to help you become familiar with the operation of PopGenLab.

Assignment 1: Getting to Know PopGenLab: Testing the Hardy-Weinberg Theorem

This assignment is designed to help you become familiar with the operation of PopGenLab by applying the Hardy-Weinberg theorem to learn about Hardy-Weinberg equilibrium parameters. The first screen that appears in PopGenLab shows you an input parameter page with a table listing the default parameters for the laboratory conditions that you can manipulate when setting up your experiments.

Before you can set up any experiment in PopGenLab, you must be familiar with the input parameters that you can manipulate for this population of moths. You will study the effects of these different parameters in future assignments. A brief description of each input parameter is provided below. Refer back to this section as needed when you are working on different assignments.

Click on the Change Inputs button to see all the parameters you can manipulate for this lab. A new page will open with buttons for each of the input parameters located at the left side of each page (genotype frequency will be open as the first input parameter). Click on each input parameter and read the descriptions below. Change each parameter so that you can become familiar with how each input parameter operates.

Genotype Frequency – allele frequency for wing color (controlled by a single gene) is shown for two alleles, A and a, as a pie chart of the phenotype resulting from each of the three possible genotypes: AA (black moths), Aa (brown moths), and aa (white moths). Notice that the default frequency is 50% for each allele.

Tree Types – three different types of trees are available. These trees provide habitat and food for the moths. Each tree has a different color of bark, which corresponds to the color of each of the three moths. Moth survival from birds as natural predators depends on a moth's ability to blend against tree bark with a color similar to the moth's wings. Notice that the default value provides an equal proportion of each tree type.

Number of Stands – the number of groups (stands) of trees determines the number of moth populations. The default value is one tree stand (one moth population).

Stand Size – the number of trees in an individual stand. Stand size directly influences the number of moths that can live within the stand. The maximum number of moths that the stand can support over time is called the carrying capacity of the stand. The default value is set to accommodate a large population, 4000 moths. This is the largest carrying capacity available in the simulation.

Migration Rate – controls the rate of migration and mixing of populations between different tree stands. *Note:* The Migration Rate button is shaded and inactive as the default because you must select at least two different populations (tree stands) for this input parameter to be active.

Mating Pattern – the type of mating can be selected as either assortative (individuals select partners with phenotypes similar to their own), random, or disassortative (individuals select partners with different phenotypes). The default is random mating. You can alter the degree of assortative or disassortative mating.

Disaster Frequency – controls the frequency of tornadoes, hurricanes, and the weather effects that typically accompany these natural disasters. Any population of organisms can be influenced by environmental disasters such as drought, flooding, hurricanes, tornadoes, and disease. In PopGenLab, the default value is for a benign environment with no disasters (Never). This frequency can be changed to produce occasional disasters (Sometimes) or frequent disasters (Often).

After you have finished this introduction to the input parameters, click the Reset button at the left of the screen to return all input parameters to their default values.

1. Testing the Hardy-Weinberg Principle: *Now that you are familiar with the basic parameters in PopGenLab, set up the following experiment to help you understand Hardy-Weinberg equilibrium and the factors that influence a state of equilibrium in a population.*

   Leave all input parameters at their default values. Check the genotype frequency value and note the allele frequency values for comparison after you have run the experiment. You now have an experimental paradigm with equal allele frequencies; genotype frequencies of 50% brown, 25% white, 25% black; equal proportions of each tree type; one tree stand with a carrying capacity of 4000 (this is a large population of moths); random mating; and because the disaster frequency is set at "Never," the population size of these moths will remain constant.

   Based on these conditions, does each moth genotype have an equal chance of survival or is there a selective pressure (positive or negative) on one or more genotypes? Explain your answer.

   Return to the initial page of the simulation if you are still in the Change Inputs view. Find the popup menu, showing the number of generations, in the lower left corner of the screen. Leave generation number set at its default value of 100. Click the Run Experiment button to run this simulation.

   When the experiment has finished running, a separate page will appear that presents the results of your experiment. Note: Any of the following data views can be saved to disk or printed by clicking the Export button. Clicking on this

button will open a separate window with your plot or table. From this window you can then save your data to your hard drive or a disk, and you can print your data by using the print feature of your browser software.
The following data can be examined:

Genotype Frequency – click and drag the slider to show genotype frequency as the frequency of AA (black), Aa (brown), and aa (white) genotypes for each generation. Data is represented as a pie chart showing the percentage of each different colored moth in the total population for each generation.

Allele Frequency – click on the check boxes to show a plotted view of allele frequency for each allele (A or a). Data is represented as the percentage of an allele in each generation.

Heterozygosity – plots the percentage of heterozygous moths in each generation over time.

Population Size – plots the number of moths in each generation over time.

Allele Distribution – plots the frequency of the A allele vs. the number of populations in each generation over time.

Textual Data – text columns of raw data for allele frequencies.

Input Summary – a summary table of the input parameters for the experiment that you carried out.

Look at the allele and genotype frequencies. Is there a change in allele frequency over time? Is there a change in genotype frequency over time? What did you observe? Explain your answers.

How do the actual genotype frequencies compare to the Hardy-Weinberg predictions?

Repeat this experiment at least three or four times to determine if the results from this experiment are consistent.

Repeat this experiment with a different set of initial genotype frequencies, but before you run the experiment, use the Hardy-Weinberg equation to predict what will happen to this population.

What did you discover? Did the results agree with your predictions? Why or why not?

2. Extending Hardy-Weinberg Principles (for use by advanced students):
Repeat the experiment described above and use Chi-square analysis with one degree of freedom to test your prediction.

How many generations does it take for a population to reach Hardy-Weinberg equilibrium?

Can different initial genotype frequencies lead to the same Hardy-Weinberg equilibrium?

Assignment 2: Genetic Drift

*Random changes in the genotypes of a population are known as **genetic drift**. Genetic drift has the potential to change the genetic composition of populations. Carrying capacity of an environment can influence how a population changes in response to genetic drift because carrying capacity is an important factor that determines the size of a population. Large populations rarely show strong effects of genetic drift; however, genetic drift in a small population can have a great influence on allele and genotype frequencies in that population. The following exercise is designed to help you understand important aspects of genetic drift as a potential cause of microevolution.*

Influence of Carrying Capacity
Formulate a hypothesis to predict the effect of a decrease in tree stand carrying capacity on allele and genotype frequencies.

Set up an experiment with default values for all of the Hardy-Weinberg conditions. Run this experiment and make note of what happens to allele and genotype frequencies (especially if you did not do assignment 1).

Run a series of experiments with default parameters for all conditions except tree stand size. For each experiment, reduce tree stand size by one-half (for example, run an experiment at 2000, half of the default value, then change tree stand size to 1000, and so on) After each experiment, make notes of what you observed for allele and genotype frequencies. Be sure to repeat each experiment several times. After you have completed this series of experiments, answer the following questions.

What happened to allele and genotype frequencies in this experiment? How do these results compare with what you observed for allele and genotype frequencies under the default conditions? Explain these results.

What happened to the percentage of heterozygotes? Explain these results.

In the following experiment, you will study several populations in different isolated tree stands.

Set up an experiment with the number of tree stands set to the maximum value of 100 and perform a series of experiments with different tree stand carrying capacities. Study genotype and allele frequencies that result from each experiment. When you are examining genotype and allele frequencies, and heterozygosity, use the text box or arrows in the lower right corner to change your view so you can see the results for the population in each tree stand as well as the average results. For each experiment, go to the Allele Distribution view and use the slider to examine allele frequency among the populations. In addition, examine what happens to population size. For each experiment, answer the following questions.

What happened to the average genotype frequencies? What happened to the genotype frequencies in each population? What happened to the allele frequencies in each population? What happened to the average allele frequencies? Explain these results.

What happened to the average heterozygosity? What happened to the heterozygosity in each population? Explain these results.

How does population size (carrying capacity) affect the changes in genotype and allele frequencies that you observed in these experiments?

What do these experiments tell you about the effect of population size on genetic variation?

Did any of the populations that you studied become extinct? If so, which ones? Why did these populations become extinct while others avoided extinction? What can you say about the relationships between carrying capacity and the risk of extinction?

Assignment 3: The Influence of Mating Patterns on Population Genetics

*One of the conditions required for maintaining Hardy-Weinberg equilibrium in a population is **random mating**. When individuals select mates with a particular heritable trait—for example, color or size—this is a form of nonrandom mating known as **assortative** mating. Similarly, inbreeding is another form of nonrandom mating in which closely related individuals within a population mate. In this experiment, you will examine how different forms of mating among moths influence Hardy-Weinberg equilibrium.*

1. Assortative Mating

   Set up an experiment at default conditions for Hardy-Weinberg equilibrium for all parameters except the number of tree stands. Set tree stand number to 100. Carry out the first experiment with random mating. Then conduct a number of experiments where there is assortative mating. Use the slider to choose different degrees of assortative mating in values ranging from 0% (random mating) to 100% (only like phenotypes mate).

   What is the effect of assortative mating on genotype frequencies? Allele frequencies? Heterozygosity? Explain your answers.

Compare the effects of assortative mating to those of genetic drift (you will need to carry out additional experiments to complete this assignment). Carry out experiments similar to those that you did for genetic drift, but change mating to different degrees of assortative mating.

How do the effects of assortative mating compare to those created by genetic drift? In particular, are the effects of assortative mating and the effects of genetic drift the same at different population sizes? Why or why not? Explain any differences you encountered between population size and the effects of assortative mating, and population size and the effects of genetic drift.

2. Disassortative Mating
    Conduct a series of experiments where there is disassortative mating. Choose different degrees of disassortative mating at values between 0% (random mating) and 100% (only unlike phenotypes mate).

    What is the effect of disassortative mating?

3. Genotype Frequencies and Mating Style
    For both assortative and disassortative mating, conduct more experiments where you vary the initial genotype frequency. Try experiments where the initial allele frequency is not equal to 0.5.

    Based on the results of these experiments, can you draw any other conclusions about the effects of assortative and dissassortative mating on allele frequency?

Assignment 4: Modes of Natural Selection

*It is possible for natural selection to affect allele and genotype frequencies in several ways depending on the mode of natural selection occurring within the population. Different modes of natural selection can select for or against a particular trait, thus affecting the balance of phenotypes in the population. For example, in a mode of natural selection called **directional selection**, selection favors individuals with one of the extreme phenotypes—a phenotype at either end of the range of phenotypes. For example, all black or all white moths are at the extremes while brown moths represent an intermediate phenotype. In **stabilizing selection**, extreme phenotypes are selected against and intermediate phenotypes have higher rates of reproduction or survival. Conversely, **diversifying (disruptive) selection** favors individuals with a range of extreme phenotypes over individuals with an intermediate phenotype.*

*In the following experiment you will investigate how **fitness**—the probability that a particular phenotype will survive and produce offspring (which is a measure of survival and reproduction of different genotypes)—affects changes in allele frequency in the population.*

Begin an experiment with default conditions for Hardy-Weinberg equilibrium for all parameters except the number of tree stands and genotype frequencies. Set the number of stands to 100. (Recall that the survival of the three types of moths depends on the distribution of tree types within the stand). Change genotype frequency to set up several different experiments under conditions of directional selection for dark color, directional selection for light color, balancing selection, and diversifying selection. For each experiment, look at genotype and allele frequencies.

Try experiments with the different conditions of selection and initial allele frequencies near zero and one.

What happens to allele frequency in the case of directional selection? Does directional selection for dark color produce the same result as directional selection for light color?

What happens to allele frequency in the case of balancing selection?

What happens to allele frequency in the case of diversifying selection?

Are small differences in fitness effective in changing allele frequencies? (Conduct additional experiments with varying proportions of tree types to help answer this question.)

Under what mode of selection might genetic variation be maintained? Explain your answer.

Under what mode of selection might there be a "founder effect," where the final allele frequency depends on the initial allele frequency? Explain your answer.

Assignment 5: Migration

Random genotypic and phenotypic changes in isolated populations can be overcome by migration. Both movement of organisms into a population and the migration of organisms out of a population can influence genotype frequencies. Migration can result in new alleles being introduced into a population. Conversely, when organisms migrate out of a population the frequency of certain alleles in the population can decline.

Run an experiment with a large amount of genetic drift occurring in 100 isolated populations by setting the number of stands to 100. To create migration among these populations, click on the Migration button and use the slider to change migration rate. Design and carry out experiments where there are different levels of migration occurring among the populations. For each of these experiments, carefully study what has happened to the populations by analyzing each category of results (genotype frequency, allele frequency, etc.).

Does migration modify the effects of genetic drift? If so, how? Explain.

Assignment 6: Population Bottlenecks

*Occasionally populations undergo "crashes," when the population size gets too small. For example, a rapid reduction in population size can occur due to natural disasters such as flood, fire, tornadoes, drought, and other extreme weather conditions. These natural disasters are frequently unselective in nature—they kill individuals throughout the population and are not selective for a particular phenotype. Even if population numbers recover, the effect of a natural disaster can have an impact on the gene pool in a population for many generations because the range of genotypes (both frequency and number) in the population that survived the disaster may not be the same as it was in the original population. Biologists call this effect a **bottleneck effect**.*

Conduct an experiment in which there is very little genetic drift. Repeat the experiment with moderate and high frequencies of "disasters" that reduce population numbers to small values.

What is the impact of these population bottlenecks?

How do these disasters affect the probability of population extinction?

Can these effects be mitigated by allowing migration to occur among the population? Yes or no? Design and carry out experiments to examine this idea. Based on your results from these experiments, if you were a conservation biologist, what would these data suggest to you about the design of natural reserves?

Assignment 7: Group Assignment

The experiments that you have conducted so far were designed to help you learn about the many parameters that influence population genetics. Typically, each of the experiments you set up analyzed the effects of a single parameter; however, it is important to realize that in nature, population genetics for any population is typically affected by changes in <u>multiple</u> parameters at the same time. The following experiments are designed to help you study the effects of different combinations of factors. Work together in a group of four students to complete these exercises.

Divide your group into pairs. For each pair of students, design and carry out a set of experiments with a different combination of parameters—for example, different rates of migration with different modes of selection, different modes of selection with different types of mating, and so on. Consider changing parameters to create situations that act in opposite directions (e.g., balanced selection with assortative mating). Before you run each experiment, formulate a hypothesis to predict the effects of your experimental conditions. Discuss this hypothesis with the other pair of students in your group and reformulate a hypothesis if necessary. Run the experiment, then analyze your results. Be sure to repeat each experiment at least twice to confirm your results.

Did the results of your experiment support your hypothesis? If not, reformulate a new hypothesis and carry out additional experiments. Explain how the experimental conditions you set up were responsible for your results. Once you think you have an explanation for your findings, design and carry out additional experiments to support or refute your thesis.

Discuss the results of your experiment with your instructor to clarify questions that you may have about interpreting the results of your experiment and understanding the effects of the experimental conditions you created.

LeafLab

Background

The Earth receives approximately 13×10^{23} calories of light energy per year from the sun. Less than 1% of this energy is captured and used by living organisms, yet without this energy, life on Earth as we know it cannot exist. Plants capture the light energy from the sun and convert it to organic molecules such as carbohydrates, in the process called **photosynthesis**. Because plants can produce organic molecules to feed themselves and support their metabolism without eating other organisms, they are referred to as **autotrophs** or **producers**. In addition to light energy, autotrophs also rely on carbon dioxide, water, and soil nutrients to produce organic molecules by photosynthesis. The summary equation showing the yield of products created by photosynthesis is shown below.

$$\text{Light Energy} + 6\ CO_2 + 12\ H_2O \rightarrow C_6H_{12}O_6 + 6\ CO_2 + 6\ H_2O$$

Other autotrophs include algae, certain protists, such as *Euglena*, and some photosynthetic bacteria. Plants are essential producers of energy for many animals, including humans. Members of the Animal Kingdom are known as **heterotrophs** because they must obtain their energy by eating other organisms. Plants are an excellent source of carbohydrates for heterotrophs. Of equal importance, plants also provide the oxygen necessary for heterotrophs to convert carbohydrates into ATP during the reactions of **aerobic cellular respiration**. It is important to remember that photosynthesis in plants is not a replacement reaction for cell respiration. Plants must still perform the reactions of cellular respiration to produce ATP; however, photosynthesis provides plants with their own source of carbohydrates. Approximately 50% of the carbohydrates produced by most plant cells are used to produce ATP via cell respiration.

The production of carbohydrates by photosynthesis can be grouped into two major metabolic stages: (1) **the light reactions** and (2) the **Calvin cycle**, also known as the dark reactions or light-independent reactions. In plant cells, both sets of reactions occur within chloroplasts. Similar to the way the reactions of aerobic cellular respiration rely on oxidation-reduction reactions, electrons produced during photosynthesis are transferred to an electron acceptor molecule called **nicotinamide adenine dinucleotide phosphate ($NADP^+$)**. Upon receiving electrons, $NADP^+$ is reduced to NADPH. Reduced NADPH functions as an electron carrier in a manner similar to the way NADH functions to supply electrons to the electron transport chain in mitochondria to power ATP synthesis in oxidative phosphorylation. The reactions of photosynthesis are summarized below.

The light reactions of photosynthesis occur in the **thylakoids** of a plant's **chloroplasts**. These reactions are designed to provide the plant cell with the NADPH and ATP required for the carbohydrate-producing reactions of the Calvin cycle. As their name indicates, the light reactions cannot occur without light energy. Specifically, most plants have evolved to absorb blue (approximately 480 nm) and red (approximately 680 nm) wavelengths of light, which are part of **visible light** spectrum. Thylakoids can absorb these wavelengths of light because embedded in the thylakoid membrane are a number of photosynthetic pigments that are capable of absorbing visible light. **Chlorophyll** *a*, which absorbs blue and red light, is the predominant pigment in the thylakoid. In addition to chlorophyll *a*, other pigments—including chlorophyll *b* and a group of pigments known as the **carotenoids**—are capable of absorbing other colors of visible light. The overall color of most plant leaves is green because the chlorophyll *a* in the chloroplasts of these leaves reflects green light.

The light reactions begin when light energy strikes chlorophyll *a* and the other pigments in the thylakoid membrane, resulting in photoexcitation of these pigments. In photoexcitation, when light energy strikes the pigment molecules, some of the electrons in these molecules are elevated to higher electron shells. These excited electrons can be captured and used by the plant cell to drive the light reactions. Electron flow in the chloroplast can occur because the photosynthetic pigments are organized in the thylakoid membrane as units called **photosystems**. Each photosystem consists of a single molecule of chlorophyll *a* called the reaction center. The reaction-center chlorophyll *a* molecule channels excited electrons to a molecule called the primary electron acceptor. Other pigments surround the reaction-center chlorophyll molecule. The pigments act as "antennae" pigments that send their excited electrons to the reaction-center molecule of chlorophyll *a*. Two types of photosystems are important for the light reactions: photosystem I and photosystem II. Photosystem I utilizes a reaction-center molecule of chlorophyll *a* called chlorophyll *a* P700, while photosystem II relies on a reaction-center molecule of chlorophyll *a* called P680. These molecules are designed to absorb visible light with a wavelength of 700 nm and 680 nm, respectively.

Excited electrons may follow one of two paths during the light reactions, **noncyclic electron flow** or **cyclic electron flow**. Noncyclic flow involves photosystem I and photosystem II. This path is the primary route for the majority of excited electrons released during the light reactions. Noncyclic electron flow begins when light energy strikes photosystem II and excited electrons from this photosystem are captured by the primary electron acceptor molecule. When this occurs, these electrons are subsequently transferred to a series of electron acceptor molecules in the thylakoid membrane. These molecules form an electron acceptor chain and many of them are very similar to those found in the electron transport chain used in cell respiration (e.g., the cytochromes). As was the case with cell respiration, electron transport along this chain results in the production of a H^+ gradient (**proton-motive force**) in the intermembrane space of the chloroplast.

As electrons are transferred along this chain, the electron carrier molecules pump hydrogen ions into the intermembrane space to create the H^+ gradient. This H^+ gradient provides the energy necessary for the enzyme **ATP synthase**, which performs the same function in photosynthesis and cell respiration. ATP synthase functions as an ion channel to allow H^+ flow down a gradient from the intermembrane space into the stroma. The H^+ flow through ATP synthase activates the enzyme to synthesize ATP from ADP and inorganic phosphate in a final stage called noncyclic photophosphorylation.

In the last part of noncyclic flow, excited electrons that have passed through the electron transport chain are now transferred to the reaction-center chlorophyll *a* P700 molecule in photosystem I. These electrons are ultimately transferred to $NADP^+$ by the enzyme $NADP^+$ reductase. This reaction results in the production of the reduced electron carrier NADPH. This entire set of reactions is called noncyclic electron flow because excited electrons that leave photosystem II never return, or cycle back, to photosystem II. How, then, does this photosystem continue to function? Why don't the photosystem molecules exhaust their supply of electrons? The answer is that the electrons in the photosystem are replaced through a water cleavage reaction that involves the enzymatic splitting of water with the concomitant removal of two electrons from water and the release of one oxygen atom. Because water splitting is constantly occurring, oxygen atoms liberated in this reaction quickly combine to form the molecular oxygen (O_2) that animals rely on to support cell respiration. The electrons released from the splitting of water are then used to replace excited electrons that leave chlorophyll *a*. Consider this reaction the next time you wonder why all your plants are dying due to a lack of water!

Although the reactions of noncyclic electron flow represent the predominant path of electrons during the light reactions, the reactions of **cyclic electron flow** are also important. The sole purpose of these reactions is to synthesize additional ATP, because noncyclic electron flow alone does not produce enough ATP to support the reactions of the Calvin cycle. This pathway involves photosystem I only. In cyclic electron flow, light strikes photosystem I and excited electrons from chlorophyll *a* P680 are transferred to a primary electron acceptor molecule. These electrons then travel down an electron transport chain. Similar to the events of cell respiration and the electron transport chain, the transfer of electrons to acceptor molecules in this chain results in the production of a H^+ gradient that is used to power ATP synthesis by ATP synthase. However, unlike the reactions of noncyclic flow, during cyclic flow the electrons that leave the electron transport chain return (hence the name cyclic flow) to photosystem I; thus the splitting of water is not required to supply electrons to this pathway. The synthesis of ATP during cyclic electron flow is called cyclic photophosphorylation.

The aforementioned light reactions are designed to supply the reactions of the Calvin cycle with the ATP and NADPH necessary for the Calvin cycle. Because the reactions of the Calvin cycle do not directly require light energy, these reactions are known as the dark reactions of photosynthesis. The primary purpose of the Calvin cycle is carbon fixation—the conversion of carbon dioxide into organic molecules such as carbohydrates. This cycle is named after Melvin Calvin, an American biochemist who was awarded a Noble Prize in 1961.

The reactions of the Calvin cycle occur in the **stroma** of the chloroplast and involve three enzymatic steps to convert carbon dioxide into valuable carbohydrates for the plant cell. Although one molecule of carbon dioxide is converted into carbohydrates during each round of the cycle, it is convenient to follow the fixation of three molecules of CO_2 (or three rounds of the cycle). In the first reaction, one molecule of CO_2 is attached to a five-carbon sugar called ribulose bisphosphate (RuBp) to produce two molecules of a three-carbon sugar called 3-phosphoglycerate. For every three molecules of CO_2 that enter the cycle, six molecules of 3-phosphoglycerate are produced. This reaction is catalyzed by an enzyme called **rubisco (ribulose carboxylase)**. The six phosphoglycerate molecules are then phosphorylated using ATP from the light reactions to create six molecules of the three-carbon sugar 1,3-bisphosphoglycerate. The six molecules of 1,3-bisphosphoglycerate are then reduced, using NADPH from the light reactions, to generate six molecules of the three-carbon sugar glyceraldehyde 3-phosphate, abbreviated G3P. G3P is an important molecule because it can be converted by the plant cell into glucose intermediates that can be used to synthesize starch and other macromolecules for the cell.

The reactions of the Calvin cycle are not completed by the synthesis of G3P, however. Of the six molecules of G3P produced by three rounds of the cycle, five molecules are used to replenish the supply of RuBp in the plant cell. This is accomplished by phosphorylating G3P in another enzymatic reaction. Therefore, of the six molecules of G3P produced, only one molecule is available as a source of consumable energy for the cell. But because these reactions occur continually, the level of G3P (and the level of carbohydrates in a cell) can accumulate rapidly. It has been estimated that worldwide production of carbohydrates by photosynthesis produces approximately 160 million metric tons of carbohydrate per year—a very large cube of sugar indeed! Approximately 50% of the carbohydrates produced by most plant cells are consumed by the cell during cell respiration. The remaining carbohydrates may be stored as starch in various parts of the plant or used to make other necessary molecules. You may have heard that talking to plants helps them grow. You should now understand why: the CO_2 that you are exhaling (and not the mellifluous sounds of your voice) promotes plant growth!

Because many plants convert carbon dioxide into the three-carbon sugar RuBp during the Calvin cycle, these plants are called C_3 plants. Not all plants perform C_3 photosynthesis. Some plants are called C_4 photosynthetic plants because, prior to the Calvin cycle, carbon dioxide in these plants is enzymatically converted into a four-carbon molecule called malate. C_4 plants include members of the grass family, such as crabgrass, and important agricultural plants such as sugarcane and corn. This reaction occurs in mesophyll cells and then the malate is shuttled into the chloroplasts of specialized cells, called bundle sheath cells, where the Calvin cycle occurs. Carbon dioxide is released from malate in the first step of the Calvin cycle, afterwhich carbon fixation proceeds according to the reactions of a C_3 plant.

What is the advantage of C_4 photosynthesis compared with C_3 photosynthesis? To understand this, consider what happens to a C_3 plant when the plant is exposed to excess heat. At high temperatures, the stomata on most plants will partially close to prevent dehydration. But while this response prevents excessive evaporation of water through the stomata, it also restricts the amount of O_2 that can leave the leaf and it limits the amount of CO_2 that can enter the leaf. Under these conditions, the elevated levels of O_2 can compete with CO_2 for binding to the active site of rubisco. When O_2 binds to the active site of rubisco instead of CO_2, rubisco attaches O_2 to RuBp to create a wasteful molecule called glycolate, which is useless to plant cells. The production of glycolate is called **photorespiration**.

The C_4 plants have adapted to avoid photorespiration when the climate is warm. The light reactions and the conversion of CO_2 into malate occur in a different set of cells (mesophyll cells) from the ones where the Calvin cycle occurs (bundle sheath cells). In addition, the attachment of CO_2 to malate requires an enzyme called PEP carboxylase. This enzyme has a higher affinity for CO_2 compared with O_2; therefore, even under conditions where the level of CO_2 in a cell is low relative to O_2, PEP carboxylase will produce malate. Because the light reactions (and the release of O_2) are occurring in mesophyll cells, when malate is pumped into the bundle sheath cells and CO_2 is released, the concentration of CO_2 in the bundle sheath cells is high enough to prevent photorespiration by rubisco. The rate of photosynthesis in both C_3 and C_4 plants can be determined experimentally by measuring the amount of CO_2 consumed by a plant's leaf. Measuring photosynthetic rate in a variety of different leaves is a primary purpose of LeafLab.

A number of different photosynthetic adaptations also occur in plants that prefer shade compared with plants that prefer direct sun. Shade-tolerant plants (e.g., ferns) often grow in the dim sunlight of a forest floor while sun plants (e.g., marigolds) prefer direct exposure to sun. Shade and sun plants have developed a number of special adaptations in response to light exposure. These include differences in photosynthetic enzymes and differences in leaf structure. Because of these adaptations, photosynthetic rate and other parameters of photosynthesis can differ in sun and shade plants when they are exposed to the same light intensity. For certain plants, both sun and shade leaves can be found on the same plant. You can use LeafLab to learn about photosynthetic rates in shade and sun plants.

Another aspect of plant biology that you will investigate using LeafLab involves the influence of plant genetics, specifically chromosome number, on photosynthetic rates. Approximately half of all flowering plant species are polyploid with respect to their chromosome number. **Polyploidy** is the presence of more than two complete sets of chromosomes in an organism's somatic cells. Unlike in animals, where polyploidy typically leads to spontaneous loss of an embryo, polyploidy in plants is partly responsible for the diversity of phenotypes in flowering plants.

Polyploidy in plants occurs when gametes contain the same number of chromosomes as somatic cells. These gametes are sometimes called **unreduced gametes** and they can arise due to **nondisjunction** of chromosomes during meiosis. Recall that during gamete formation by meiosis, the number of chromosomes is typically reduced to half so that gametes have a haploid (n) number of chromosomes compared with diploid ($2n$) somatic cells. Unreduced gametes can lead to a number of different polyploidy conditions in plants, including triploid ($3n$; typically sterile plants) and tetraploid ($4n$) plants. Hexaploids ($6n$) and octaploids ($8n$) are also fairly common, and several plant species have been studied with significantly higher numbers of chromosomes. Plant breeders sometimes induce conditions of polyploidy to create new plant species. Examples include certain ryes and fruits, such as seedless watermelons and bananas.

When studying photosynthesis, botanists are routinely interested in learning about many different aspects of photosynthesis in addition to photosynthetic rate. Processes such as dark respiration, photosynthetic saturation, photochemical efficiency, and light compensation points are other important measures of photosynthesis. LeafLab can be used to study all of these processes. You will use LeafLab to learn about many of the factors presented in this background section by simulating experiments that modern-day botanists use to study the environmental and genetic factors that influence the rate of photosynthesis in plant leaves.

References

1. Mauseth, J. D. *Botany: An Introduction to Plant Biology*, 2nd ed. Sudbury, MA: Jones and Bartlett, 1998.

2. Raven, P. H., Evert, R. F., and Eichhorn, S. E. *Biology of Plants*, 6th ed. New York: Freeman, 1999.

3. United States Department of Agriculture Natural Resources Conservation Service Web Site. http://plants.usda.gov/

Introduction

In this laboratory, you will perform simulations of experiments designed to study the reactions of photosynthesis as they occur in the leaves of different plants. Some of these plants perform C_3 or C_4 photosynthesis, some plants prefer shade over direct sunlight, while some plants in LeafLab have different numbers of chromosomes that will affect photosynthetic rates. By changing experimental parameters such as light intensity, light quality, temperature, gas flow, and carbon dioxide concentration, you will learn about the importance of each parameter by measuring the amount of carbon dioxide consumed by the plant cells in your experiment as they undergo the reactions of photosynthesis. Data collected from these experiments will be calculated to determine photosynthetic rates.

Objectives

The purpose of this laboratory is to:
- Demonstrate how photosynthetic rates in different plants can change in response to factors such as light intensity, light quality, CO_2 concentration, and temperature.
- Simulate measurements of CO_2 assimilation rates in leaves.
- Investigate dark respiration, photochemical efficiency, CO_2 conductance, carboxylation efficiency, light compensation points, and photosynthetic saturation.
- Compare photosynthesis in C_3 and C_4 plants.
- Study the effects of polyploidy on photosynthetic rates.

Before You Begin: Prerequisites

Before beginning LeafLab you should be familiar with the following concepts:
- The importance and functions of enzymes as biological catalysts, basic principles of metabolic pathways, and mechanisms involved in regulating the catalytic activity of an enzyme (see Campbell, N. A., Reece, J. B., and Mitchell, L. G. *Biology* 5/e, and Campbell, N. A., and Reece J. B., *Biology* 6/e, chapter 6).
- The structure and function of the chloroplast (chapters 7 and 10).

- The electromagnetic spectrum and the photoexcitation of chlorophyll by visible light (chapter 10).
- The reactions of photosynthesis including cyclic and noncyclic electron flow in the light reactions, and the Calvin cycle. Be able to describe the primary substrates required, reactions involved, and products generated by each of these reactions (chapter 10).
- Plotting, interpreting data, and fitting data points to curves in scatter plots; using y- and x-intercepts, slope of the line, and the asymptote to extrapolate data from a line plot.

Assignments

For your ease in completing each assignment, the background text relevant to the experiment that you will perform is italicized, instructions for each assignment are indicated by plain text, and questions or activities that you will be asked to answer are indicated by **bold** text.

The following assignment is designed to help you become familiar with the operation of LeafLab.

Assignment 1: Getting to Know LeafLab: Measuring Photosynthetic Rate in Tomato Plants

The first screen that appears in LeafLab takes you to a virtual lab containing all of the equipment you will need to carry out your experiments.

Click on each piece of equipment to learn about its purpose.

For each experiment you conduct, you must understand the experimental setup you will be using and manipulating. The basic experimental design begins with a lamp as a source of visible light. The intensity of visible light produced by this lamp can be increased or decreased, and the wavelengths of visible light released by the lamp can be altered using different filters that will allow only certain wavelengths of visible light to strike the leaf you are studying. To prevent the leaf sample from drying out or burning due to heat from the lamp, a reservoir of water is placed between the lamp and the leaf. LeafLab allows you to choose from several different leaf samples.

The leaf is contained in a sealed chamber. In addition to manipulating the quantity and quality of light striking the leaf in the chamber, you can manipulate various environmental conditions of the leaf chamber, such as gas flow, temperature, and CO_2 concentration. When collecting data, the concentration of CO_2 in the leaf chamber will be measured using an infrared gas analyzer (IRGA). Because water vapors that can affect the accuracy of the IRGA will be produced in the chamber when light strikes the leaf, air leaving the chamber is passed through a drying column prior to entering the IRGA. The IRGA measures the amount of infrared radiation absorbed by CO_2 in the air coming from the leaf chamber. Analyzing data on the amount of CO_2 consumed by your leaf will enable you to study several different parameters of photosynthesis.

Are you comfortable with the purpose of each piece of equipment in LeafLab? Be sure that you understand the experimental design before continuing with this assignment.

1. *In this first experiment, we will consider the effect of light intensity on photosynthetic rate in tomato leaves.*

    Based on what you already know about photosynthesis, develop a testable hypothesis to explain the influence of an increase in light intensity on the photosynthetic rate in tomato leaves.

    a. *Choosing a Leaf and Measuring Leaf Surface Area*
    To set up each experiment, first choose the leaf you are interested in studying. Leaves from six different plants are available in LeafLab. The plants available to you are tomato (C_3 plant), corn (C_4 plant), two different clones of goldenrod (one favors sunny conditions, the other favors shade), and two types of fescue grass that differ by their number of chromosomes.

    Click on the Choose Leaf button on the left side of the screen. For this experiment we will use a leaf from a tomato plant. A dark box should appear around the tomato leaf in the left corner of this screen indicating that the tomato plant has already been selected (tomato is the default plant). Read the legend about tomato plants that appears on the right side of the screen. Similar legends will appear for each plant that you select. Also notice that the bottom three panels of this screen show graphics of the whole plant, leaf, and fruit structures for the tomato plant. Similar legends and graphics will appear for each plant you select. Review some of the other plants available in LeafLab by clicking on each plant button and reading the legends that accompany each plant.

    Because the calculations of photosynthetic rate that this simulation will generate as data are expressed as a value per unit of leaf surface area, you must begin each experiment by determining the total surface area of the leaf you have chosen for your experiment

    Click on the Measure Area button on the left side of the screen. A view of a tomato leaf will appear, overlaid by a series of grid squares. Click on one of the squares. It will now be shaded green, indicating that you have measured this area of the leaf.

    A tally of the number of squares selected and the total surface area (in cm^2) for all squares selected is provided on the right side of the screen. The scale for measuring area is different for certain plants.

    Notice that the scale for tomato plants is set up so that each square is 0.1 cm^2 in area. Continue to select squares until you have measured the entire surface area of this leaf; however, when you get to the edges of the leaf, notice that some of the grid boxes are not completely filled by the leaf. For these boxes, double-click on each box. The box will now be shaded a lighter green than the other boxes. Notice that these boxes are scored as one-half of the area of the other boxes to adjust your area measurement for incompletely filled boxes. <u>Hint</u>: You can measure surface area quickly by using your mouse to click and drag across several boxes.

The surface area value that you just measured should be recorded in your lab notebook for future reference as follows:

Click on the Add to Notes button at the lower right of the screen to record this value in your notebook.

For each plant in LeafLab, only one size leaf will appear; therefore, once you have measured leaf area for a particular plant you will not have to measure the area of this leaf again unless you exit LeafLab and then return again. The area measurement will be available to you even if you switch plants to perform another experiment. If you do not measure leaf surface area at the beginning of an experiment, LeafLab will not let you run the experiment. Once you have defined the total surface area of the leaf, you are almost ready to run an experiment; however, before you can begin collecting data you must know what environmental parameters in the leaf chamber you can manipulate and the types of data you will be collecting.

Click on the Collect Data button on the left side of the screen. The screen that appears is labeled Input Controls. For this first experiment, we will leave many of the Input Controls at their default value. At the bottom of the screen, locate the box labeled "Expt #." This box will identify each experiment that you perform so you can return to this experiment if you need to. This first experiment should be indicated by a "1" in this box.

b. <u>Input Controls</u>
The Input Controls view allows you to change several conditions in the chamber. These include gas flow into the leaf chamber, temperature, carbon dioxide concentration, light intensity, and light quality. Gas flow is measured in milliliters of gas entering the chamber per minute (ml/min). The default value for gas flow is "off." Temperature is measured in degrees Celsius (°C). The default value for temperature is set at 25°C, the temperature of a warm room (80.6°F), which is a fairly comfortable temperature for most plants living in a typical greenhouse. Carbon dioxide concentration is measured in parts per million (ppm). The default value is 350.0 ppm. This value approximates the normal atmospheric concentration of carbon dioxide. Light intensity is measured in micromoles of light photons released per square meter per second ($\mu mol/m^2/s$). The default value is that the light is turned off (0 $\mu mol/m^2/s$). Unless you are performing an experiment to measure photosynthetic rates under darkness, you must always adjust light intensity before you can collect data. The light filters control allows you to determine which wavelengths of light you would like the leaf to be exposed to. The default is no filter; thus, white (visible) light is striking the leaf.

112

Find the input controls for each of the parameters mentioned in the preceding paragraph, then set up the conditions for this experiment as follows:

Set gas flow to a medium value by clicking on the "medium" button. Notice that gas flow has now changed from 0 ml/min to 500 ml/min.

Gas flow is a parameter that you will change depending on the size of the leaf you are working with. For tomato and goldenrod leaves, medium gas flow is appropriate. For corn, a high gas flow is necessary. For fescue, the small blades of grass require a low gas flow. If you try to run an experiment without gas flowing into the leaf chamber, no data will be generated when the lamp is turned on.

Leave temperature, carbon dioxide concentration, and the light filter at their default values. Unfiltered white light will be shining on the tomato leaf when we turn on the lamp intensity. It is best to turn on either the lamp or the gas flow as the last step in the experiment. Begin this experiment by leaving lamp intensity at 0 μmol/m^2/s. Note: An intermediate value of approximately 1000 μmol/m^2/s is close to representing a typical value of sunlight on a sunny day. Temperature, CO_2 concentration, and light intensity can be changed either by using the slider for each parameter or by entering a value into the text box that appears to the right of each slider.

Notice that once you turn on the lamp, a chart recording of CO_2 output, as determined by the IRGA, will appear in a box just above the chart recording.

Locate the numerical value for CO_2 output in the leaf chamber.

The IRGA is determining CO_2 output by comparing the amount of CO_2 entering the chamber with the amount of CO_2 in the air leaving the chamber. Remember that the leaf is consuming CO_2 as its cells perform photosynthesis but these cells are also undergoing cell respiration to produce ATP—a process that produces CO_2 as a waste product. Therefore, the IRGA is recording CO_2 output as a measure of <u>net photosynthetic rate</u>—the difference between CO_2 consumption during photosynthesis and CO_2 production during cell respiration. <u>Note</u>: It is important to realize that you are not measuring O_2 production in these experiments.

To take measurements of CO_2 output, click the Record button in the bottom right corner of the screen. When you do this, data on CO_2 output will be recorded in tabular form in the Prepare Data view of LeafLab. Before recording any measurements, always make sure that the line on the chart recording is horizontal and not wavy before taking a measurement. This is especially important after you have changed an input parameter—because you must wait for the experimental conditions in the leaf chamber to "settle down" to your desired input setting(s).

Click on the Record button to take a recording of CO_2 output. Notice that when you click the Record button, a solid black line appears on the chart recording, indicating that a measurement was taken.

Continue this experiment by increasing light intensity by increments of approximately 200 μmol/m^2/s to 200, 400, 600, 800, 1000, 1200, 1400, 1600, 1800, and 2000 μmol/m^2/s. It is easy to make these changes by typing these values into the text box to the right of the light intensity slider. After each change in light intensity, wait for the chart recording to flatten out to a horizontal line and then record your data by clicking the Record button.

c. *AnalyzingData*
   To prepare your data for analysis, switch to the Prepare Data view.

   Switch to the Prepare Data view by clicking on the Prepare Data button at the left of the screen.

The purpose of the Prepare Data function is to use the data that you recorded to calculate photosynthetic rate. In this view, the table at the bottom of the screen contains values for all of the input parameters of each experiment that you conducted. This table also indicates the concentration of CO_2 entering the leaf chamber (C-in) and the concentration of CO_2 leaving the leaf chamber (C-out). The calculated value for photosynthetic rate (P) will appear in the far right column of the table. LeafLab will calculate P for us in a moment.

The four equations at the top of this screen will be used to calculate photosynthetic rate. Each equation is described below.

Equation 1: Calculates CO_2 consumption by subtracting C-out from C-in and reports this value as the change in CO_2 concentration (ΔCO_2).

Equation 2: Converts ΔCO_2 concentration from parts per million to μmol/liter, based on the temperature of the leaf flask.

Equation 3: Calculates the rate of CO_2 that is available for exchange between the leaf and the flask by multiplying ΔCO_2 concentration against gas flow.

Equation 4: Net photosynthetic rate (P) is calculated by dividing the CO_2 available for exchange by the total surface area of the leaf being studied. This value is reported as the number of micromoles of CO_2 released per square meter of leaf surface area per second (μmol/m^2/s).

   Look at the table and review the information contained in each column before calculating photosynthetic rate.

114

To enter data into these equations, click on the row for the experiment that you want to analyze—in this case, experiment 1. The selected row will now be highlighted in green. Select the whole table by clicking on the top row (0 light data) and dragging the mouse to the bottom of the table. To perform the calculations, click on the Compute button, located on the right side of the screen just above the P column in the table. Net photosynthetic rate will now be calculated and added to the table.

d. _Plotting Data_
Plotting data from the calculations that you generate is important for understanding the results of your experiments. To do this, use the Plot Data function of LeafLab. This function will produce a scatter plot of your data.

Click on the Plot Data button on the left side of the screen. Click in the title box at the top of this screen and title this first plot "PS Rates vs. Light in Tomato" or another appropriate name of your choice.

You can express your data on these plots in two different ways. On the x-axis you can choose to plot either "light intensity" (this is the default) or CO_2 input (C-in). On the y-axis you will plot photosynthetic rate (P) values. You can also change the symbol and symbol color that will appear on the plot. Prepare a plot of light intensity versus P values as follows:

Start at the top row of data in the table, and plot the entire table of P values by holding down the Shift key and clicking on each row of data until you get to the last row of the table. The entire table should now be colored green, indicating that you selected all of the data in the table. Leave the x-axis at its default value of light intensity. Click the Plot Selected Data button to plot the data. A plot of your data should appear. Notice that the horizontal line at the center of the plot indicates a "0" value of light intensity.

Add the data from this plot to your notebook by clicking on the Add to Notes button at the lower left of the screen.

e. _Interpreting Your Data Plots_
Although basic trends in your data can sometimes be estimated by simply looking at the data points on your scatter plots, quantitative measures of the effects you are studying can only be determined by fitting a curve to your data. Curve fitting is an important part of studying correlations between data plotted on a scatter plot. Curve fitting involves producing a statistically derived best fit of data points and not a hand-drawn or estimated line connecting data points. LeafLab uses a statistical technique called least squares for estimating a best-fit line. See your instructor for additional help if you are uncomfortable or unfamiliar with curve fitting.

Once you have plotted your data, a "Plot #" tab will appear at the top of the Plot Data screen. Clicking on this tab will take you to the curve-fitting functions of LeafLab and allow you to switch between plots that you generate.

Click on the Plot 1 tab to enter the curve-fitting view.

An enlarged view of the plot should now appear with a series of curve-fitting controls to the left of the plot. The purpose of each control is described below.

<u>Curve</u> *– generates a best-fit curve based on the data points selected.*

<u>y -Intercept</u> *– indicates the rate of dark respiration (light compensation point)*

<u>Slope</u> *(of the line) –* **photochemical efficiency***; the rate at which photosynthesis increases as light intensity increases.*

<u>Asymptote</u> *(where the curve forms a straight line indicating that the data has leveled off) – indicates* **photosynthetic saturation** *(maximum rate of photosynthesis).*

<u>Error SS</u> *– error sum of squares; based on the calculations of the least squares parameter estimation of a best-fit line. Error SS is an expression of the least squares calculation of the (sum and squared) distances of each data point from the fit line.*

When fitting a curve it is helpful to proceed as follows:

(1) Change the intercept first. To do this, return to the data table by clicking on the Data tab. Look at the zero light measurement in the table and use the P value for this measurement as the initial measurement of the intercept. Return to the curve-fitting view and enter this P value directly into the intercept box.

(2) Look at the plot and form a rough estimate of where the data levels off. This is the asymptote. To determine this value, click on the plot next to the data point that you think represents the asymptote. Two sets of numbers in parentheses will appear. The first number is light intensity (in nanometers) and the second number is photosynthetic rate (P). Enter this P value into the asymptote box.

(3) Next, increase the slope of the line until the curve looks like it is matching (fitting) the data points. Do this by clicking the up arrow next to the slope function (you will see the line rise up and begin to form a curve). Make adjustments to the slope of the curve and asymptote as necessary to achieve the best-fit curve for the data points.

(4) Look at the value for the Error SS. Adjust the slope, asymptote, and intercept to minimize this number. Use the up or down arrows next to

each parameter to adjust these values. If you make a change to one parameter, you will need to check the other two parameters to see if further changes are necessary. Stop only when any further changes to all three parameters increase the Error SS value. The values that give the smallest Error SS (the least squares parameter estimates) produce the best-fit line for your data points.

(5) Save your plot by clicking on the Export Graph button at the left of the screen. A separate window will now open showing your plot and a table with the intercept, slope, asymptote, and Error SS values. You can print this page by clicking the print button on your Web browser, or you can save this page to a disk by going to File and using the Save As feature of your browser.

f. *Summary: What Did This Experiment Tell Us?*
The experiment you just performed is representative of other experiments that you will conduct. A lot of information can be learned from studying the curves that you generate. Seek help from your instructor if you are unfamiliar with how to study a curve to interpret values on the x- and y-axis. Study the curve of PS Rates vs. Light in Tomato to answer the following questions (Hint: You may need to refer back to section e for definitions of each parameter of the plot):

What is the relationship between an increase in light intensity and photosynthetic rate in tomato leaves? Does this relationship support the hypothesis that you formulated?

Photosynthetic saturation is the maximum rate of photosynthesis. What was the value for photosynthetic saturation in tomato leaves? What value of light intensity produced photosynthetic saturation in tomato leaves? Based on what you know about photosynthesis, provide possible reasons for what causes photosynthetic saturation (these cannot be determined from the plot).

2. Effects of Light Intensity on Photosynthetic Rates in Corn
Follow the steps detailed in the first experiment to test the effects of an increase in light intensity on photosynthetic rates in corn (a C_4 plant). The only modification to the experiment is that you will need to use a high rate of gas flow. Keep all other parameters the same as you did for tomato. (Note: When calculating P and plotting your data, make sure that you select only those values that you recorded for corn and not previously recorded values for tomato.) Plot photosynthetic rate versus light intensity and fit a curve to the data as you did for tomato, then answer the following questions:

What is the relationship between an increase in light intensity and photosynthetic rate in leaves from a corn plant? How does this relationship compare with what you observed for tomato plants?

Photosynthetic saturation is the maximum rate of photosynthesis. What value of light intensity produced photosynthetic saturation in corn leaves?

Assignment 2: Influence of Light Quality on Photosynthesis

The visible light spectrum consists of many colors of light of different wavelengths, ranging from 380 nm to 750 nm; however, not all colors of light are equally effective at stimulating photosynthesis in plant leaves. This assignment is designed to investigate the influence of light quality on photosynthesis in corn.

Based on what you already know about photosynthesis, which colors of visible light are most effective for photosynthesis in plants? Which colors of visible light are least effective for photosynthesis? Why are some colors more effective than other colors?

Develop a hypothesis to predict the effect of changing light quality from white light to red, green, and blue light on photosynthetic rates in corn.

Test your hypothesis as follows:
Choose corn and set the CO_2 level to the default value, gas flow to high, and temperature to 25°C. Set the light filter to white and measure photosynthesis for light values of 0, 200, 400, 600, 800, 1000, 1200, 1400, 1600, 1800, and 2000 $\mu mol/m^2/s$.

Repeat this experiment using red, green, and blue light filters. Remember that the light shining on the leaf corresponds to the filter that you have chosen.

Prepare and fit separate curves of photosynthetic rate (P) versus light intensity for the data from each experiment using the different light filters. (Hint: You cannot get a good asymptote estimate for any filter except white light; therefore, use the asymptote estimate for white light and fit only the intercept and slope.)

What effect does light quality have on the rate of photosynthesis? Which colors of light were most effective at stimulating photosynthesis? Can you relate your results to the absorption spectra of the photosynthetic pigments?

Assignment 3: Comparing C₃ and C₄ Plants

*In the first assignment, you had the opportunity to study photosynthesis in a C₃ plant (tomato) and a C₄ plant (corn). Recall that C₄ plants have developed adaptations to avoid **photorespiration**, a condition that will affect C₃ plants when the climate is hot. This assignment is designed to help you compare differences between C₃ and C₄ plants using the data that you generated in assignment 1.*

Begin by creating a third plot of both your corn and tomato data from assignment 1. To do this, return to the Plot Data view. Name this plot "Plot 3" by entering a "3" in the Plot box. Select the data for tomato and plot this data.

Change the value in the Data for Curve box to "2" (Note: This box indicates the number of different data curves that you will plot on the same graph). Select the corn data. Choose a different plotting symbol and/or color for the corn data, then plot the corn data on the same graph with the tomato data.

Click on the Plot 3 tab to enter the parameter values that you fit previously to tomato and corn to develop best-fit curves for the data. Answer the following questions.

> **Are there any differences between photosynthetic rate in corn compared with tomatoes? If there are differences, what are they? How do C_3 and C_4 plants differ in their capacity for photosynthesis? Are there differences in photosynthetic saturation in corn compared with tomatoes? What are they? Provide possible explanations for any differences that you observed.**

Assignment 4: Who Can Stand the Heat? Sun Versus Shade Plants
As you learned in the background text for this lab, photosynthetic rate in sun and shade plants differs in response to light intensity. The following experiment is designed to examine and compare photosynthetic rate in a sun clone of goldenrod with that of a shade clone of goldenrod.

> Repeat the same experiment as with tomato and corn using goldenrod sun and shade clones. Leave all input parameters at their default values except for gas flow; turn this to medium. Plot data for both plants on the same graph and fit curves to each, then answer the following questions:

> **What are the adaptations of these two clones for growing in different light conditions? Why is the photosynthetic saturation point different in these two clones? Are these results what you would have expected based on what you know about sun plants and shade plants? Explain your answer.**

Assignment 5: Light Compensation Points
Although plants are consuming CO_2 during the dark reactions of photosynthesis, they are also producing CO_2 as they produce ATP during cell respiration. Light compensation occurs when the rate of photosynthesis balances the rate of cell respiration such that the net rate of CO_2 production is zero. In this experiment, you will learn to measure light compensation points in two different clones of goldenrod.

> Perform an experiment using goldenrod sun and shade clones. Gas flow should be on medium. All other parameters can remain at their default values. For each leaf, measure photosynthetic rates for light intensities of 0, 20, 40, 60, 80, 100, and 120 $\mu mol/m^2/s$.

Plot the data for both leaves on the same graph. Enter the asymptote values from your previous plots of the goldenrod (assignment 5) and fit the intercept and slope, keeping the asymptotes constant. Note the point at which the curves cross the x-axis. This is called the light compensation point. Click and drag the mouse on the plot to determine the exact value for the light compensation point.

What was the value of the light compensation point for each of these clones? If you were simply measuring net CO_2 concentration for a leaf at its light compensation point, you could interpret the data to mean that no photosynthesis was occurring in the leaf. Explain why this would not be a correct interpretation.

Do the differences in light compensation points between sun and shade clones make sense given the conditions under which these clones would be growing in their natural ecosystems?

Assignment 6: Effect of CO_2 Concentration on Photosynthesis

The amount of available CO_2 in a leaf can strongly influence the rate of photosynthesis. One measure of the ambient (circulating) CO_2 concentration available within the leaf is called carboxylation efficiency. Carboxylation efficiency is a measure of CO_2 movement into and around the leaf (leaf conductance) compared with CO_2 transfer to ribulose bisphosphate during the first step of the Calvin cycle—a reaction called carboxylation.

In the first exercise of this assignment you will study carboxylation efficiency in tomato (C_3 plant); in the second exercise of this assignment you will study carboxylation efficiency in corn (C_4 plant). At the end of the second exercise, you will be asked to answer several questions about carboxylation efficiency in C_3 and C_4 plants.

1. Set up an experiment with tomato and set light intensity to 1000 μmol/m²/s. Leave temperature at its default value of 25°C, set gas flow to medium, and use the white light filter. Begin with a CO_2 concentration of 0 ppm and continue this experiment by increasing CO_2 concentration to 100, 200, 300, and so on, until you reach 1000 ppm. Record CO_2 output after each change in CO_2 concentration. Plot photosynthetic rate (P) versus C-in and fit a curve to the data.

   Interpret your plots as follows:
       Intercept = rate of dark respiration
       Slope = carboxylation efficiency
       Asymptote = photosynthetic saturation (maximum rate of photosynthesis)
2. Repeat the experiment that you set up in exercise 1; however, use a leaf from corn and set gas flow to high. Plot P versus C-in for both tomato and corn and fit curves to both, then answer the following questions:

How does the carboxylation efficiency differ between C_3 and C_4 plants? How does photosynthetic saturation differ between C_3 and C_4 plants? Are these differences in photosynthetic saturation consistent with the differences that you observed between C_3 and C_4 plants in other experiments? Do these differences make sense based on what you know about the anatomy of C_3 and C_4 plants? Explain your answers.

Assignment 7: Effect of Temperature on Photosynthesis

Temperature is another one of the many environmental conditions that will affect photosynthesis in plants. Even if you haven't studied the effects of temperature on photosynthesis before, you are probably aware that certain plants in and around your home (plants in your vegetable garden, weeds in your lawn such as crabgrass) grow better or grow faster at certain temperatures. The following experiment is designed to help you learn about the effects of temperature on photosynthesis in tomato leaves.

Choose tomato and set the CO_2 level to the default value of 350 ppm, gas flow to medium, and the light filter to white. Set temperature to 15°C and measure photosynthesis for light values beginning at 0 $\mu mol/m^2/s$ and increasing by 200 $\mu mol/m^2/s$ until you reach maximum light intensity of 2000 $\mu mol/m^2/s$. Repeat this experiment for temperature values of 20°C, 25°C, and 30°C. On the same plot, fit curves of P versus light intensity to the data for each of the four temperatures, then answer the following questions:

What effects does temperature have on the rate of photosynthesis in tomato leaves?
How might these effects be different in a C_4 plant such as corn? Develop a testable hypothesis to explain the influence of temperature on photosynthesis in corn, and conduct experiments to confirm or refute your hypothesis.

Assignment 8: Influence of Polyploidy on Photosynthesis

You have already investigated the effects of many different environmental parameters on photosynthesis in C_3 compared with C_4 plants and sun plants compared with shade plants. However, genetic factors also influence photosynthetic rate in plants. As described in the background text, plants with extra sets of chromosomes—called polyploid plants—often have different photosynthetic properties than plants of the same species with a normal number of chromosomes. In the following exercise, you will examine the effects of polyploidy in two clones of a hardy grass called tall fescue—a very popular turfgrass. One strain of fescue is a tetraploid (4n; four sets of chromosomes) and another strain is an octaploid (8n; eight sets of chromosomes).

Repeat the experiment you did with tomato and corn in assignment 1, exercises 1 and 2 using tall fescue 4n and 8n. Leave all parameters at their default values. Set gas flow to low. Begin recording at 0 light intensity and continue this experiment by increasing light intensity by increments of approximately 200 $\mu mol/m^2/s$ to

200, 400, 600, 800, 1000, 1200, 1400, 1600, 1800, and 2000 $\mu mol/m^2/s$. Record after each change in light intensity.

Plot curves for both plants on the same graph and fit each curve.

How does polyploidy affect leaf photosynthesis? What differences did you observe? Provide possible reasons why chromosome number influences photosynthetic rate.

Assignment 9: Group Assignments

In the previous assignments, you used LeafLab to investigate the effects of several different conditions on photosynthetic rate in different leaves. One condition that affects photosynthetic rate in virtually all plants is temperature. The following exercises are designed to help you investigate the effects of temperature on all of the plants in LeafLab and to compare data for these plants. Work together in a group of four students to complete these exercises.

Divide your group into pairs and have each pair pick three plants to work on. For each plant, perform an experiment that involves four or five different temperatures, ranging from low to high temperature. Use the same range of temperatures for each plant. For each plant, make sure that you pick the ideal values for gas flow that were used in the previous assignments. Leave the CO_2 level set at its default value of 350 ppm. Measure photosynthesis for light values beginning at 0 $\mu mol/m^2/s$ and continue this experiment by increasing light concentration to 200, 400, 600, and so on, until you reach 2000 $\mu mol/m^2/s$.

For each plant, fit curves of P versus light intensity to the data for each of the temperatures that you selected. Print the plots for each plant so you can compare your data with data from the experiments conducted by the other pair of students in your group. Work together with the other students in your group to analyze your data as described below. To compare some of these data, you may need to gather your data together and draw your own plots with data from several different plants on the same plot.

For each plant, draw conclusions about the effects of temperature on photosynthetic rate, dark respiration rate, photochemical efficiency, and photosynthetic saturation.

For which plants did temperature show the greatest effects on the parameters of photosynthesis that you studied? Which plants were affected the least by changes in temperature?

Based on what you know about the properties of each plant, such as the type of photosynthesis it utilizes (C_3 versus C_4), preferred environmental conditions for growth, and locations where each plant typically grows, explain whether the temperature effects that you saw for each parameter make sense or not.

CardioLab

Background

For any animal to carry out its metabolism and survive, it must be able to exchange gases between its cells and the environment. Oxygen gas (O_2) must enter the body for the organism to perform cell respiration, and CO_2 generated as a waste product during cell respiration must be removed from the body. Gas exchange requires the transport of these molecules. For example, O_2 moves from the lungs to body tissues and CO_2 moves from body cells to the lungs so that it can be exhaled. The transport of gases and a variety of other molecules necessary for life—for example, nutrients, ions, and hormones—is accomplished by organs and tissues of the cardiovascular system. In addition, the cardiovascular system must transport metabolic waste products to facilitate their removal from the body. In mammals, the cardiovascular system consists of blood as the transport medium, the heart, and blood vessels.

Human blood is a connective tissue that consists of two major components, an aqueous fluid called **plasma** and the formed elements. Plasma makes up approximately 55% of blood volume. The formed elements consist of red blood cells (erythrocytes), white blood cells (leukocytes), and platelets. Formed elements make up approximately 45% of blood volume. Some molecules dissolve directly in the plasma, many molecules must be attached to transport molecules in the plasma to facilitate their movement through the circulatory system, and others such as oxygen gas are transported within red blood cells. Regardless of how a molecule is transported within an animal's blood, the organism's survival requires a pump to move blood through the circulatory system. Of course, the heart is the sophisticated pump that accomplishes this purpose.

The human heart is a four-chambered structure located in the thoracic body cavity. The two upper chambers of the heart are called the **atria** (singular, atrium), and the two lower chambers are called **ventricles**. When the heart contracts, both atria contract almost simultaneously, followed by contraction of both ventricles. Before we discuss the events that trigger contraction of the heart, it is important that you understand how blood flows to, through, and away from the heart. Oxygen–poor, carbon dioxide–rich blood enters the right atrium through two large veins, the superior vena cava and the inferior vena cava. The venae cavae collect oxygen–poor blood from virtually all veins in the body. As the heart contracts, blood that has entered the right atrium flows into the right ventricle then from the right ventricle into a large artery called the pulmonary trunk. The pulmonary trunk carries blood away from the heart into one right and one left pulmonary artery. The pulmonary arteries serve to carry oxygen–poor blood to the vessels of the lungs wherein carbon dioxide and oxygen are exchanged. Following oxygenation in the lungs, blood enters the left atrium via four pulmonary veins. During each contraction of the heart, blood flows from the left atrium to the left ventricle. From the left ventricle, blood enters the largest artery in the body, the aorta. The aorta gives rise to major arteries of the body that transport oxygen–rich blood to all body organs. The right side of the heart and the blood vessels that carry blood to and from the lungs form the pulmonary circuit. The left side of the heart, the aorta, and vessels derived from the aorta are part of the systemic circuit because these structures deliver oxygen–rich blood to body organs.

Another important aspect of heart anatomy involves connective tissue valves that function to ensure the one-way flow of blood through the heart. One set of valves, the semilunar valves (aortic and pulmonary), function to prevent the backflow of blood into the ventricles. Two atrioventricular valves, the tricuspid valve and the bicuspid (mitral) valve, prevent the backflow of blood into the right and left atrium, respectively, when the ventricles contract.

Biologists use the term *cardiac cycle* to describe one complete sequence of events involved in pumping and filling the heart with blood. In one complete beat of the heart, the atria contract nearly simultaneously, followed by contraction of the ventricles. Technically, contraction of the heart chambers is known as **systole** while relaxation of the heart chambers is called **diastole**. When the atria are in systole, the ventricles are in diastole; therefore, both ventricles fill with blood. Alternatively, when the ventricles are in systole, the atria are in diastole; this is when the atria fill with blood.

The frequency and duration of the cardiac cycle is determined by the electrical events that govern cardiac muscle contraction. However, unlike skeletal muscle, which will only contract in response to electrical impulses from motor neurons in the brain, cardiac muscle cells contract in response to electrical impulses that are generated within the heart itself. Because these impulses arise from within the heart, the generation and spreading of electrical currents in the heart is known as the intrinsic conduction system. The intrinsic conduction system is so effective that a human heart can continue to beat on its own if removed from the body! Although neurons from the brain are not responsible for causing the heart to beat, neurons that are part of the autonomic nervous system (ANS) are very important for regulating the frequency of heartbeats and contraction strength.

The intrinsic conduction system consists of groups of specialized muscle cells called conduction fibers. Some cells of this system simply conduct electrical impulses to other regions of the heart while others, known as pacemaker cells, spontaneously produce electrical impulses (also called action potentials) that determine heart rate. The intrinsic conduction sequence of electrical events is initiated by a small cluster (node) of cells located in the right atrium called the sinoatrial (SA) node. The SA node is considered the true "pacemaker" of the heart because it produces impulses at a rate of approximately 75 impulses/minute. This node of cells determines the overall rate of contraction by all other cells in the heart. As electrical impulses from the SA node spread to cardiac muscle cells in the atria, those cells are stimulated to contract (atrial systole). The impulse moves from cells of the atria to a node of cells, called the atrioventricular (AV) node, located between the two ventricles. The impulse is delayed momentarily by the AV node. This delay allows the ventricles to fill with blood while the atria are in systole. After this delay, electrical impulses travel through a series of cells called the atrioventricular bundle (bundle of His) to cells called Purkinje fibers. The Purkinje fibers are located in the walls of the ventricles; therefore, once electrical impulses have reached the Purkinje fibers, the ventricles contract (systole). This entire sequence takes approximately 0.2 seconds. The aforementioned electrical events are measured and recorded using an electrocardiograph, which produces a tracing of these events called an electrocardiogram (ECG or EKG).

Neurons of the ANS can regulate heart rate by influencing the pacemaker cells, particularly cells of the SA node. Sympathetic nervous system neurons release the neurotransmitter norepinephrine, which triggers SA node cells to produce electrical impulses with greater frequency, thus increasing heart rate. Epinephrine released by the adrenal gland also increases heart rate. Exercise and stress—for example, going on a blind date or being anxious about a biology exam—can cause an increase in heart rate by stimulating sympathetic nervous system neurons. This increase in heart rate is designed to provide the additional oxygen and nutrients that body organs require to increase their metabolism and respond to the stress. This response is commonly referred to as the fight-or-flight response. Conversely, neurons of the parasympathetic nervous system oppose the actions of the sympathetic nervous system. Parasympathetic neurons release acetylcholine, a neurotransmitter that decreases the frequency of impulse generation by cells of the SA node. These parasympathetic effects are responsible for returning heart rate to normal after a stressful event. The parasympathetic nervous system also slows heart rate during times when body organs have a reduced requirement for oxygen, such as when you are sleeping. Thus, the ANS is critical for the fine-tuned control of heart rate necessary to ensure the proper transport of nutrients to body organs. Other details about how the ANS regulates heart rate and blood vessel diameter are discussed later in this section.

Although the heart is a pump in the simplest sense, the factors that control heart rate are multidimensional and complex. Physiologists use the term *hemodynamics* to describe the study of blood flow, blood pressure, and the regulation of those processes necessary for blood pressure homeostasis. One very important concept in hemodynamics is **cardiac output (CO)**. Cardiac output is the amount of blood pumped by one ventricle in one minute (ml blood/minute), and is determined by the following relationship:

$$CO = SV \text{ (stroke volume)} \times HR \text{ (heart rate)}$$

Stroke volume (SV) is the amount of blood (in milliliters) pumped by one ventricle during one beat. In a healthy human heart, SV is approximately 70 ml. Stroke volume is a measure of the difference between end diastolic volume and end systolic volume in each ventricle. End diastolic volume is the amount of blood that fills each ventricle at the end of ventricular diastole, approximately 120 ml/ventricle. End systolic volume is the amount of blood that remains in each ventricle at the end of ventricular systole. Although the heart is a fairly efficient pump, it is impossible to expel all blood from a ventricle when it contracts. End systolic volume is approximately 50 ml of blood per ventricle. Thus, each contraction, or "stroke," of a ventricle expels approximately 70 ml of blood. At rest, normal **heart rate (HR)** rate is approximately 75 beats/minute. Using these values, CO at rest is approximately 5250 ml of blood per ventricle per minute. During exercise, stress, or in response to a number of stimuli, CO can increase dramatically to accommodate an increased need to supply body tissues with oxygenated blood. In fact, in well-conditioned athletes, CO can increase by more than four times the normal value of 5250 ml/minute!

Normal CO is required to deliver the amount of blood necessary to maintain normal body functions. One benefit of exercise is that most forms of exercise—particularly high-impact aerobic exercise—can lead to an increase in the elasticity of the heart, which means that the ventricles are capable of filling with more blood. This increase in elasticity produces an increase in stroke volume. Because stroke volume increases in a well-conditioned athlete, resting heart rate can decrease and an athlete can still continue to maintain CO—this is just one beneficial aspect of exercise. Conversely, individuals who do not exercise their heart on a regular basis can show less elasticity in the wall of the ventricles and thus a lower stroke volume. Therefore, to maintain normal cardiac output, heart rate must compensate by increasing—thus putting more strain on the heart. When one considers that this individual may compound a lack of exercise with other conditions such as a diet rich in saturated fats and cholesterol—both of which can accumulate in the chambers of the heart—it is easy to understand why a lack of exercise and poor diet are considered risk factors for cardiovascular disease.

Normal cardiac output is necessary for maintaining proper blood pressure—technically defined as arterial pressure—the amount of pressure blood exerts on the wall of an artery. The normal blood pressure value of 120/80 recorded when you visit your physician is a measure, in millimeters of mercury (mm Hg), of systolic blood pressure created by contraction of the left ventricle compared with diastolic pressure during ventricular diastole.

Blood pressure will rise with ventricular systole and fall with ventricular diastole. This rise and fall can be measured as a "pulse" on the wall of an artery as elastic tissue in the arterial wall stretches during ventricular systole and recoils during ventricular diastole. It is particularly important that pressure within the cardiovascular system be sufficient to keep blood moving through the system properly. Mean arterial pressure (MAP) is the average blood pressure within systemic arteries. This pressure represents the average pressure due to ventricular systole and is considered the "driving force" that delivers blood to body organs.

Another important measure of hemodynamics is blood flow. Blood flow refers to the volume of blood moving through an area over a given time. Typically, flow is expressed in milliliters of blood per minute. Blood flow is affected by two major factors: blood pressure and total peripheral resistance. The relationship of these factors is shown below.

$$\text{Blood Flow (F)} = \frac{\text{Blood Pressure (BP)}}{\text{Total Peripheral Resistance (TPR)}}$$

Blood flow is directly proportional to blood pressure and inversely proportional to resistance. In this relationship, blood pressure actually refers to the difference in pressure between two points—for example the two ends of a blood vessel. Resistance refers to any factor that interferes with or opposes blood flow. Total peripheral resistance is a measure of all resistance factors that blood encounters as it travels through the cardiovascular system. It is important that you be familiar with the many factors that contribute to resistance. Blood vessel diameter, blood vessel radius, and blood thickness (viscosity) are particularly important factors. There is an inverse relationship between blood vessel diameter and blood resistance. A large–diameter blood vessel has less resistance than a small–diameter blood vessel because in a larger blood vessel, there is less friction between blood cells themselves and the walls of the vessel. A good example of this is the difference between drinking from a drinking straw and drinking from a thin-diameter straw used to stir coffee. The thin-diameter straw has a much higher resistance. This is also the reason firefighters use large–diameter fire hoses instead of small–diameter garden hoses.

Technically, resistance is inversely proportional to vessel radius to the fourth power. Therefore, water traveling in a hose with a 1-inch radius experiences 16 times more resistance than blood traveling through a hose with a 2-inch radius. This same principle also applies to blood vessels. As a result, very small changes in vessel diameter can produce large changes in peripheral resistance, blood flow, and blood pressure. Resistance is also directly proportional to vessel length. The longer the path that blood must travel, the greater the friction between blood and the vessel wall. Therefore, blood traveling in shorter blood vessels encounters less resistance than blood traveling through longer vessels. Although both vessel diameter and length influence blood pressure and flow, vessel diameter exerts a much greater influence over hemodynamics than vessel length. This is because vessels in humans rarely change in length; however, vessel diameter can and does change a lot in response to conditions such as exercise, stress, and temperature changes. An increase in blood vessel diameter is called vasodilation, while a decrease in vessel diameter is known as vasoconstriction.

In addition to regulating heart rate directly, the ANS is also involved in controlling vasoconstriction and vasodilation. For example, during prolonged exercise the ANS will trigger vasodilation of blood vessels in skin and muscles to improve blood flow to muscles. This process also functions to cool body temperature by radiating heat away from the skin. Conversely, vasoconstriction can occur during conditions of stress as a way to increase blood pressure.

The ANS cannot control heart rate or blood vessel diameter without the involvement of sensory neurons that signal the ANS to produce necessary changes. An important set of sensory neurons called baroreceptors are located in the wall of the aorta and the internal carotid arteries. These neurons respond to changes in blood pressure and relay impulses to the medulla oblongata of the brain. In the medulla, impulses are relayed to clusters of neurons that form important cardiovascular control centers. One group of neurons, called the cardioacceleratory center, consists of sympathetic nervous system neurons that, when stimulated, send electrical impulses to the SA and AV nodes to increase heart rate. Related neurons also send impulses to smooth muscle cells in the walls of blood vessels to stimulate vasoconstriction. Therefore, if blood pressure were to decrease for any reason, baroreceptors would send impulses to the brain that stimulate sympathetic nervous system neurons to increase heart rate and trigger vasoconstriction in an effort to raise blood pressure. Conversely, when blood pressure increases, baroreceptors send impulses to the medulla to inhibit neurons of the cardioacceleratory center. In this situation, the baroreceptors also stimulate a group of neurons in the medulla called the cardioinhibitory center. These neurons are part of the parasympathetic division of the ANS, and when stimulated they send impulses to the SA and AV nodes to decrease heart rate. Most blood vessels do not receive impulses from parasympathetic neurons; hence, vasodilation is brought about by the inhibition of sympathetic neurons. This neural control is essential for blood pressure homeostasis.

Another important factor that influences blood pressure is blood viscosity. Blood viscosity will change in conditions such as anemia, where a lack of blood cells decreases blood thickness. When viscosity decreases, blood flows with less resistance, thus decreasing blood pressure. Similarly, decreases in blood volume due to excessive bleeding (hemorrhaging) will also reduce blood pressure. Conversely, blood viscosity can increase when an individual is producing an excess number of red blood cells (a condition called polycythemia) and when an individual is experiencing excessive dehydration. "Thicker blood" produces an increase in peripheral resistance.

Clearly, blood pressure homeostasis involves many aspects of cardiovascular physiology. Familiarity with the relationships between blood pressure, blood flow, resistance, and cardiac output is essential for understanding and appreciating the complexities of hemodynamics. It is possible to study basic aspects of hemodynamics in human patients by following changes in cardiovascular functions during exercise, stress, and other conditions. However, it is obviously impossible to perform invasive experimental manipulations on human patients for the benefit of helping you learn about the human cardiovascular system and blood pressure homeostasis. CardioLab will provide you with an outstanding opportunity to learn about important parameters that influence blood pressure and blood flow. You will be manipulating parameters such as heart rate, blood vessel radius, blood viscosity, ventricular volume, venous capacity, and blood volume to simulate the effects of changes in these parameters on hemodynamics. Through these experiments you will learn how organs of the cardiovascular system maintain blood pressure homeostasis in response to different stimuli. You will also use CardioLab to simulate real–life conditions that affect blood pressure such as exercise, hemorrhaging, shock, and cardiovascular disorders.

References
1. Marieb, E. N. *Human Anatomy and Physiology*, 4th ed. Menlo Park, CA: Benjamin/Cummings, 1998.

2. Silverthorn, D. U. *Human Physiology: An Integrated Approach*, 1st ed. Upper Saddle River, NJ: Prentice Hall, 1998.

Introduction

In this laboratory, you will perform simulations of experiments designed to study the relationships between the various parameters that influence blood pressure and heart rate in humans, and examine how the cardiovascular system responds and functions to maintain blood pressure homeostasis.

Objectives

The purpose of this laboratory is to:

- Demonstrate important relationships between heart rate, stroke volume, resistance, cardiac output, and mean arterial pressure.
- Investigate how changes in various parameters of the cardiovascular system such as vessel radius, ventricular volume, blood viscosity, and heart rate affect blood pressure and how blood pressure homeostasis is restored in response to these changes.
- Simulate the effects of chemicals, shock, hemorrhage, exercise, and cardiovascular disorders on blood pressure homeostasis in humans.

Before You Begin: Prerequisites

Before beginning CardioLab you should be familiar with the following concepts:

- The organization and functions of the mammalian cardiovascular system and the structure and function of the human heart (see Campbell, N. A., Reece, J. B., and Mitchell, L. G. *Biology*, 5/e, and Campbell, N. A., and Reece J. B., *Biology* 6/e, chapter 42).
- The composition of human blood (chapter 42)
- Physiological events in the human heart, including the intrinsic conduction system, the cardiac cycle, and the pathway of blood flow to, through, and away from the heart (chapter 42).
- Principles of hemodynamics, including relationships between blood flow, blood pressure, and peripheral resistance (chapter 42).
- Autonomic nervous system regulation of heart rate, blood vessel diameter, and blood pressure homeostasis (chapter 48).

Assignments

For your ease in completing each assignment, the background text relevant to the experiment that you will perform is in brown text, instructions for each assignment are indicated by plain text, and questions or activities that you will be asked to answer are indicated by **bold text**.

The following assignment is designed to help you become familiar with the operation of CardioLab.

Assignment 1: Getting to Know CardioLab: Factors That Affect Cardiac Output and Mean Arterial Pressure:

The first screen that appears in CardioLab presents the relationship between resistance, cardiac output (CO), and mean arterial pressure (MAP). This feature of CardioLab is known as the "Equation." It is essential that you understand the relationships of these factors before beginning an experiment. The Equation feature is designed to help you do this. Follow the exercises below to examine the effects of resistance and cardiac output on MAP. These exercises are an excellent way to reinforce your understanding of important relationships that influence MAP. You can manipulate any of the parameters in this view and watch how your change influences MAP. This feature is not designed to demonstrate homeostasis; therefore, you will not see these parameters change to return MAP to normal. Homeostasis will be investigated in the other assignments.

1. Effect of Blood Viscosity on Mean Arterial Blood Pressure

   A number of different conditions can influence blood viscosity. For example, blood viscosity will decrease due to a decrease in the number of red blood cells in the condition known as anemia. Conversely, individuals living at higher altitudes often experience polycythemia—an abnormal increase in red blood cell count. Polycythemia occurs in response to reduced oxygen content of the atmosphere at higher altitudes. Both decreases and increases in blood viscosity strongly influence MAP.

   Develop a hypothesis to predict the effect of an increase in blood viscosity on blood pressure, then test your hypothesis as follows.

   Click and drag on the slider to increase blood viscosity.

   What happened to MAP? Does this make sense to you? Explain your observations and relate them to your hypothesis.

   Use the slider to decrease blood viscosity and observe what happens to MAP.

   What happened to MAP? Does this make sense to you? Explain your observations and relate them to your hypothesis.

2. Effect of Blood Vessel Radius on Mean Arterial Pressure

   Formulate a hypothesis to predict the effect of an increase in blood vessel radius on MAP. Formulate a separate hypothesis to predict the effect of a decrease in blood vessel radius on MAP.

   Test each hypothesis by using the slider to change blood vessel radius and follow the effects of these changes on MAP.

   What happened to MAP after each change? Do these effects make sense to you? Explain your observations.

In the cardiovascular disease called arteriosclerosis ("hardening of the arteries"), the deposition of saturated fats and cholesterol along the inner lining of blood vessels reduces vessel diameter.

Simulate this condition and explain what happens to MAP.

3. Effect of Heart Rate on Cardiac Output and Mean Arterial Pressure
   Formulate a hypothesis to predict the effect of an increase in heart rate on cardiac output and MAP. Formulate a separate hypothesis to predict the effect of a decrease in heart rate on cardiac output and MAP.

   Test each hypothesis by using the slider to change heart rate and follow the effects of these changes on cardiac output and MAP.

   What happened to cardiac output after each change? What happened to MAP after each change? Do these effects make sense to you? Explain your observations.

4. Effect of Stroke Volume on Cardiac Output and Mean Arterial Pressure
   Increase stroke volume by increasing diastolic ventricular volume, then observe what happens to cardiac output and MAP.

   Explain why increasing diastolic ventricular volume produced an increase in stroke volume. What happened to cardiac output and MAP when stroke volume was increased?

   Based on what you know about stroke volume, what is another way that stroke volume can be increased using CardioLab?

   Once you have answered this, use CardioLab to verify or refute your answer.

   Are you comfortable with the basic relationships between MAP and resistance? Be sure that you understand the relationships between MAP, resistance, and cardiac output before continuing with these assignments.

Assignment 2: Getting to Know CardioLab: Blood Pressure Homeostasis
Once you have a comfortable understanding of the relationships between MAP, cardiac output (CO), and resistance, you can use CardioLab to perform experiments that will help you understand the many mechanisms involved in blood pressure homeostasis.

> Click the To Experiment button at the lower left corner of the screen to leave the Equation view.

The Variables view that will now appear provides you with many options for designing an experiment. A number of different conditions (variables) can be manipulated using this feature to help you understand blood pressure homeostasis and the relationships of hemodynamics. Notice that you can use sliders to manipulate heart rate, vessel radius, blood viscosity, systolic ventricle volume, blood volume, and venous capacity—the amount of blood contained within systemic veins.

You can manipulate these variables to study the effects of each variable on important measures of cardiovascular system physiology and to demonstrate how the human body controls many different aspects of the cardiovascular system during homeostasis. When you run an experiment in which you have changed any of these variables, the simulation will ultimately return each variable to normal (default values), allowing you to examine how each variable responds to the experimental manipulation that you created. For each experiment, you are provided with output data using chart recordings for five important measures of cardiovascular activity and hemodynamics. These include the following:

Mean Arterial Pressure (mm Hg)
Heart Rate (beats/min)
Stroke Volume (ml)
Total Peripheral Resistance (dyne-s/cm^5)
Blood Volume (L)

The time frame for each recording is in seconds. A numerical value for each parameter will also appear in the far right column of these recordings. Note: The numerical values for each recording can be saved in your lab notebook or printed by clicking on the Export Text button at the lower left side of the screen. A separate window will open. From this window, you can print your data using the print feature on your browser or you can save this data to a disk or to your hard drive.

Look at the other functions of CardioLab by clicking on each tab at the top of the screen.

In some experiments, you will use the Interventions feature to apply certain experimental conditions that affect the cardiovascular system. The Cases feature will be used to examine three different cardiovascular disorders, while the Nerve Impulses feature will be used to view electrical activity to and from the heart.

In this exercise you will examine the normal values for each output measure.

1. Normal Values of Cardiovascular Physiology
   You must understand what the normal values for each parameter are before you manipulate any of these parameters.

   In the Variables view, click on the Start button to begin the simulation. Carefully examine the normal values for mean arterial pressure, heart rate, stroke volume, total peripheral resistance, and blood volume. Look at both the patterns of each recording and the numerical values for each measure (shown at the far right of each recording). Be sure that you are comfortable with the normal values for each parameter before moving to the next exercise.

Click on the Nerve Impulses view to study electrical activity in a normal patient.

Notice that several different tracings are shown. Below is a description of the purpose of each recording.

Carotid Sinus – *this recording is measuring electrical activity from a cluster of neurons that are located in the wall of each internal carotid artery. These neurons are called baroreceptors because they sense blood pressure in the carotid artery and send electrical impulses to the medulla oblongata of the brain. Baroreceptors are very important for the feedback mechanisms involved in blood pressure homeostasis. Impulses from the baroreceptors are integrated in the medulla to control ANS neurons, which in turn can increase or decrease heart rate or influence blood vessel diameter according to blood pressure changes in the carotid artery.*

Vagus – *this recording is measuring electrical activity of the vagus nerves. The vagus nerves are cranial nerves that transmit approximately 75% of parasympathetic nervous system activity in the human body. Activity of the vagus nerves regulates heart rate in addition to regulating the involuntary functions of many other body organs. The vagus nerves innervate the SA and AV nodes of the heart. Electrical impulses from these nerves inhibit activity of the SA and AV nodes to decrease heart rate.*

Sympathetic Cardiac – *this recording is measuring electrical activity of the sympathetic cardiac nerves. As their name indicates, these nerves are part of the sympathetic division of the ANS. Sympathetic cardiac nerves innervate the SA and AV nodes. Electrical impulses from these nerves increase activity of the SA and AV nodes to increase heart rate.*

Sympathetic Vasoconstrictor – *this recording is measuring electrical activity of sympathetic nervous system nerves that are innervating smooth muscle cells in the walls of systemic arteries. Recall that few arteries are innervated by parasympathetic neurons. Sympathetic innervation of the arterial wall is primarily responsible for changes in the diameter of arteries. In general, stimulation of these nerves triggers vasoconstriction of systemic arteries while a decrease in electrical activity in these nerves triggers vasodilation of systemic arteries.*

Take note of the normal values for each electrical activity. In other experiments you will study how these electrical activities change in response to different conditions.

Click on the Stop button to stop this simulation, then click on the Reset All button to reset the simulation.

2. Effect of Heart Rate

In this exercise you will study the effects of a change in heart rate on other parameters of the cardiovascular system. Before you run your experiment, consider the following questions:

What effect will a change in heart rate have on each of the (five) other parameters indicated in the Variables view? Imagine that you have just walked into your biology class and your instructor has surprised you with a rather lengthy, unannounced essay exam on the cardiovascular system. Your heart rate increases in response to this stress. How will this increase in heart rate affect other parameters of the cardiovascular system such as blood pressure, stroke volume, and total peripheral resistance? Which of these other parameters will change in an effort to maintain blood pressure homeostasis? How will each parameter change? What role will the nervous system play in homeostasis?

Studying these changes will help you understand how the human body will compensate in an attempt to maintain blood pressure homeostasis. Set up an experiment to answer these questions as follows:

Click on the Start button and allow normal recordings to continue for 5 seconds. After 5 seconds, click and hold on the slider for heart rate and increase heart rate to close to the maximum value for this slider. Click on the box next to this slider to freeze heart rate at this value. Notice that you have increased heart rate.

Look at the slider bars for vessel radius, blood viscosity, systolic ventricle volume, blood volume, and venous capacity. Observe what is happening to each of these parameters. Take note of the following:

Which parameters changed and how did each parameter change? For example, what happened to vessel radius? Did vessel radius increase or decrease? Why? Do these results make sense to you?

Did any parameter(s) remain unchanged? If so, which one(s)? Do these results make sense to you? Explain your answers and relate them to understanding of blood pressure homeostasis to explain why each parameter did or did not change in response to an increase in heart rate.

Once you have answered these questions, stop the simulation and repeat this experiment. This time, look at the tracings at the bottom of the screen and take note of any changes in each output parameter.

Did mean arterial pressure return to normal? Why or why not? What happened to stroke volume? Peripheral resistance? Blood volume? Explain your answers.

Stop the stimulation and repeat this experiment. This time, look at the Nerve Impulse tracings and take note of any changes that you see.

What happened to the electrical activity of each set of nerves? Which nerves showed an increase in electrical activity? Which showed a decrease? Did electrical activity stay the same for any of these nerves? Explain your answers.

Repeat the experiment described above, but this time decrease heart rate and answer the same questions presented for the increase–in–heart–rate experiment.

3. Effect of Blood Viscosity

As you learned in the first assignment, blood viscosity can decrease due to different anemias, and increase due to polycythemia. In the first assignment, you looked at the effect of changes in blood viscosity on MAP; however, you did not study how the cardiovascular system will respond to viscosity changes in an effort to maintain blood pressure homeostasis.

Repeat the steps described in exercise 2 above; however, this time instead of changing heart rate, increase blood viscosity. Then answer the following questions:

Which variables changed to compensate for the increase in blood viscosity? Based on what you know about cardiovascular relationships and hemodynamics, do these changes make sense to you? Why or why not? Explain what you observed.

Did MAP return to normal? Why or why not? Explain your answers.

Repeat this process to study a decrease in blood viscosity.

Assignment 3: Blood Loss Compensation

The Interventions feature of CardioLab is designed to test how well you understand the principles of hemodynamics by providing you with the opportunity to design simulations of patients with various changes in hemodynamics and then to intervene in an attempt to return heart rate and blood pressure to normal. One condition that you can simulate is hemorrhage—the loss of blood. A classic, immediate symptom of hemorrhage is a decrease in mean arterial blood pressure due to a decrease in blood volume. Hemorrhage can lead to inadequate circulation of blood, a general condition known as shock. There are several different forms of shock and conditions that can lead to shock; however, one common form of shock due to the loss of blood is called hypovolemic shock.

The cardiovascular system can respond to hemorrhage quickly by activating short-term compensatory mechanisms designed to maintain blood pressure homeostasis and avoid shock. Depending on the extent of blood loss, other long-term compensatory responses involving the kidneys and bone marrow can be stimulated. The following exercises are designed to help you study the effects of blood loss on the cardiovascular system.

1. Small Hemorrhage

    How can we help a patient overcome problems due to hemorrhaging? In addition to the obvious and immediate treatment—stopping the blood loss—consider other treatments for a patient who is hemorrhaging.

    Imagine that you are a nurse or a physician caring for a woman who has just delivered a child. Childbirth proceeded without complications and this woman is showing no outward signs of hemorrhaging. She is now sleeping comfortably. Although there are no outward signs of hemorrhaging, she is experiencing a small amount of (internal) bleeding from her uterus. Simulate this condition as follows:

    Click on the Interventions tab at the top of the screen. Notice that Small Hemorrhage appears in the popup menu as the default condition. Click Start to begin the simulation, allow this to proceed normally for several seconds, and then click on the Apply Intervention button to induce a small amount of bleeding.

    Carefully look for changes in each recording. Which parameters changed? Which remain unchanged? Explain these results. Does blood pressure return to normal in this patient due to homeostasis? If so, explain which parameters changed to allow her blood pressure to return to normal.

    Repeat this experiment if necessary to observe all changes.

2. Large Hemorrhage

    While homeostasis can often compensate for small losses in blood volume, large losses of blood (for example, a 20% loss of blood volume) typically require medical interventions to return blood pressure to normal. This exercise is designed to help you understand the effects of a large loss of blood.

    A man is walking across the street in a busy city at rush hour and is accidentally struck by a delivery truck. Fortunately, a school crossing guard has noticed the accident and immediately calls for an ambulance. Upon arrival at the accident scene, the emergency medical personnel notice that this individual has been bleeding excessively from a wound to his leg. Simulate this condition as follows:

    Click on the Start button to begin the simulation. In the Interventions view, click on the popup menu and select Large Hemorrhage, then click on the Apply Intervention button.

    Notice the immediate decrease in blood volume. Note any other changes in cardiovascular parameters. How is this patient's cardiovascular system responding in an effort to achieve blood pressure homeostasis? What is happening to heart rate? Does this make sense? Why or why not? What is happening to stroke volume?

What is happening to total peripheral resistance? Explain each observation.

Do these responses return blood pressure to normal?

If not, you will need to intervene as follows. Assume that you have taken appropriate measures to stop the hemorrhaging. Help this patient by increasing his blood volume through a blood transfusion. To do this, click on the Variables tab. Click on the slider for blood volume and gradually increase blood volume while monitoring blood pressure and heart rate. Give this patient more blood until you have given him enough to return his blood pressure and heart rate to normal. Hold blood volume steady until blood pressure and heart rate have stabilized.

What happened to other parameters of the cardiovascular system as you gave him more blood? Explain these changes.

You may need to repeat this experiment again to observe all changes as they occur. Hopefully, congratulations are in order for your role in helping this patient survive and recover from his wounds! If you were unable to maintain blood pressure in this patient, repeat the experiment.

Assignment 4: Effects of Blood Volume Changes

In addition to hemorrhage, many other conditions can lead to changes in blood volume. These can include diet, hormonal imbalances, kidney disorders, and lung disorders among other conditions. These assignments are designed to help you understand the effects of increases and decreases in blood volume.

1. Increase in Blood Volume
   Imagine that you are a nurse in a hospital caring for a man with a kidney infection and kidney stones. Because of these problems, this patient is not producing sufficient amounts of urine. He is receiving nutrients and antibiotics administered through an intravenous (IV) line. You just changed his IV bag to a new bag containing 1 liter (L) of solution, but you inadvertently adjusted this IV bag so that fluid is entering the patient too rapidly.

   What may happen to this patient if the entire bag is administered? How will this increase in body fluid affect important cardiovascular parameters? Consider these questions before you begin the simulation.

   Simulate this situation by clicking on the Start button to monitor cardiovascular parameters in a normal human. Allow this to proceed for a few seconds. In the Interventions view, select IV Infusion from the popup menu, then click on the Apply Intervention button twice in rapid succession to simulate the IV infusion of a large volume of fluid.

What happens to each of the following parameters: MAP, heart rate, stroke volume, total peripheral resistance, and blood volume? Does each response make sense to you? Explain each effect.

2. Decrease in Blood Volume (No Hemorrhage)
You are well aware that exercising on a hot day can lead to excess loss of fluids via sweat. Of course, one danger of excess fluid loss is dehydration. However, you may not realize that dehydration is not just the loss of sweat. Virtually all body fluids, both intracellular fluids and extracellular fluids such as plasma, cerebrospinal fluid, interstitial fluid, and synovial fluid, are shared and mixed in a dynamic fashion designed to maintain the total body water content of approximately 40 L in humans. Because of these fluid dynamics, a decrease in fluid from one compartment, such as the loss of fluid from sweat glands, produces a decrease in most of the other fluid compartments. Hence, excess sweating decreases blood plasma volume. This assignment is designed to study the effects of dehydration on the cardiovascular system.

Formulate a working hypothesis to predict the effects of a decrease in blood volume on MAP.

Simulate this situation by clicking on the Start button to monitor cardiovascular parameters in a normal human. Allow this to proceed for a few seconds, then in the Interventions view, select Dehydration from the popup menu and click on the Apply Intervention button.

What happens to each of the following parameters: MAP, heart rate, stroke volume, total peripheral resistance, and blood volume? Does each response make sense to you? Explain each effect. Relate these observations to your hypothesis.

Assignment 5: Exercise and the Cardiovascular System
Exercise-induced effects on the cardiovascular system are largely due to changes in electrical impulses coming from neurons of the autonomic nervous system (ANS). These changes are designed to ensure adequate delivery of blood to exercising muscles. The following exercise will help you learn about the effects of exercise on the cardiovascular system.

Based on what you already know about the cardiovascular system and based on what happens to your heart rate when you exercise, think about how the ANS is responsible for exercise-induced changes. What role does the sympathetic division of the ANS play in the cardiovascular response to exercise? How does this occur? Ask yourself the same questions about the parasympathetic division of the ANS, and the role of baroreceptors. Also consider how each neural component changes when you begin to relax following exercise.

Click on the Start button to monitor cardiovascular parameters in a normal human. Allow this to proceed for a few seconds, then in the Interventions view, select

Treadmill from the popup menu and click on the Apply Intervention button four or five times in rapid succession to simulate strenuous exercise.

Notice what happens to each cardiovascular parameter. Record any changes that occur and explain these results.

To observe what is happening to nerve activity, stop this experiment by clicking the Stop button. Click the Reset All button. Repeat the experiment as before; however, before you apply the treadmill intervention, be sure to switch to the Nerve Impulses view to examine normal impulse activity. Once you have done this, apply the treadmill intervention as before, then return to the Nerve Impulses view to look at exercise–induced changes in nerve impulse activity.

What happened to impulse activity from each set of neurons? Describe your results.

Follow the activity of these neurons for several minutes until homeostasis of blood pressure and heart rate is achieved, then answer the following questions:

Describe the role of each set of neurons in blood pressure homeostasis. Which components of the cardiovascular system were affected by each group of neurons?

Assignment 6: Neural Control of the Cardiovascular System

As described in the background text for this laboratory, homeostasis of blood pressure is achieved through ANS control via two mechanisms—regulating heart rate and controlling blood vessel diameter. Neurons from both the sympathetic division and the parasympathetic division of the ANS utilize neurotransmitters to mediate these changes. Recall that neurons of the sympathetic division release epinephrine onto heart cells and smooth muscle cells in the walls of blood vessels, while neurons of the parasympathetic division rely on the neurotransmitter acetylcholine. This assignment is designed to examine the effects of both neurotransmitters on heart rate, blood vessel diameter, and resulting changes in cardiovascular parameters.

1. Fight-or-Flight Response of the Sympathetic Nervous System
   On the way home from meeting a friend for dinner, your car breaks down in a dark alley in an unfamiliar town at 1:30 a.m. The battery on your cell phone is dead, so you need to walk to the closest pay phone to call for help. As you walk down the alley you can feel the fight-or-flight response of your ANS starting to take effect as your heart rate increases, your temples throb, your pupils dilate, and you start to sweat. Simulate this situation as follows:

   Click on the Start button to monitor cardiovascular parameters in a normal human. Allow this to proceed for a few seconds. In the Interventions view, select Epinephrine from the popup menu, then click on the Apply Intervention button once. Go to the Nerve Impulses view and observe what is happening to the activity of each group of neurons. Explain your results. You may need to switch back and forth between the Nerve Impulses view and the Interventions view to explain how changes in nerve impulse activity may relate to changes in cardiovascular parameters such as heart rate and total peripheral resistance.

2. Medical Uses of Epinephrine
    Injections of epinephrine are used to treat certain patients with heart problems.

    Can you think of a condition or multiple condition(s) when it might be helpful to treat someone with epinephrine?

    Design an experiment to mimic this condition, then carry out this experiment using CardioLab.

Assignment 7: Chemical Effects on the Cardiovascular System

In addition to the neurotransmitters that regulate the cardiovascular system, a variety of other commonly used chemicals—such as nicotine and caffeine, medicinal drugs, and illicit drugs—strongly influence heart rate. One example of an illicit drug with such effects on the cardiovascular system is cocaine. You may be aware of cocaine-induced heart attacks from several widely publicized cases involving athletes and professional actors. Cocaine produces an increase in heart rate and is also a potent stimulator of vasoconstriction.

Another chemical that strongly influences heart rate is foxglove, also known as digitalis. Foxglove is derived from flowering plants in the genus Digitalis. Dried leaves from Digitalis purpurea can be prepared in powdered form as a potent source of foxglove. Foxglove is a compound that functions as a vasodilator and also as a cardiotonic drug—a drug that increases contraction strength of the heart. Foxglove is often used for the treatment of patients with a cardiovascular disorder known as congestive heart failure. You will simulate the use of foxglove on patients with congestive heart failure in the next assignment. The following assignment is designed to help you examine the effects of foxglove.

    Click on the Start button to monitor cardiovascular parameters in a normal human. Allow this to proceed for a few seconds, then in the Interventions view, select Foxglove from the popup menu and click on the Apply Intervention button once.

    What happens to heart rate and blood pressure after you give foxglove? What happens to total peripheral resistance? Explain your results.

Assignment 8: Cardiovascular Disorders

Unfortunately, cardiovascular disorders are fairly common in the United States. These assignments are designed to help you understand how three relatively common disorders influence the physiology of cardiovascular organs.

1. Congestive Heart Failure (CHF)

   It has been estimated that over 8 million people in the United States and Europe suffer from congestive heart failure. This condition is a progressive disorder characterized by a decrease in cardiac output, and consequently blood flow, due to a number of different problems. In many patients with CHF, the problem is due to an enlarged left ventricle that is weak and ineffective at adequately pumping blood into the cardiovascular system. In other patients, damage to muscle cells in the walls of the left ventricle prevent the ventricle from maintaining normal cardiac output. In either case, because the left ventricle is unable to pump with the same effectiveness as the right ventricle, blood backs up into the pulmonary circuit, creating "congestion."

   In the Cases view, start an experiment to follow normal heart rates. After a few seconds, stop the normal run by clicking the Stop button. Simulate CHF by selecting Congestive Heart Failure from the popup menu and then clicking on the Apply Case button. To begin the simulation, click the Start button. Observe what happens to each of the cardiovascular parameters in a CHF patient. Follow this patient for several seconds until you have a clear picture of which cardiovascular parameters may be changing. Note: In the Cases view, if you are already running an experiment you cannot just "give" a patient a disorder. You have to stop the simulation. Apply the case and then click Start to run the chosen case.

   Describe the symptoms that this patient is experiencing. Do they make sense to you based on what you know about CHF? Why or why not?

   In the previous exercise you studied the effects of foxglove (digitalis) and learned that it is often used to treat CHF.

   Simulate this treatment by returning to the Interventions view. Select Foxglove from the popup menu, then click on the Apply Intervention button.

   What effect does foxglove have on cardiovascular parameters in the CHF patient? Carefully study any changes in cardiovascular parameters and describe what these changes are.

2. Hypertension

   Elevated blood pressure is called hypertension. Hypertension is a complex cardiovascular disorder with a number of different causes resulting from known risk factors such as stress, gender, obesity, and elevated levels of plasma cholesterol and saturated fats. Other cases of hypertension can be attributed to hormonal imbalances, kidney problems, and associated disorders in fluid balance in the body. In addition, there are a number of cases of hypertension for which the causes remain unclear. Because the development of hypertension is often multifactorial in nature, it can sometimes be very complicated to treat. Typically, hypertension is treated by trying to minimize risk factors that can be controlled in combination with the use of different drugs that affect heart rate and blood vessel diameter.

Simulate hypertension by selecting Hypertension from the popup menu and then clicking on the Apply Case button. Observe what happens to each of the cardiovascular parameters in a hypertensive patient. Follow this patient for several seconds until you have a clear picture of which cardiovascular parameters may be changing.

> **Think about which parameters you might want to manipulate to help this patient. Because many of the drugs that are used to treat hypertension affect heart rate and blood vessel diameter, see if you can use the Variables feature to lower blood pressure in this patient.**
>
> **Discuss desirable effects that will lower blood pressure in hypertensive patients.**

3. Mitral Valve Stenosis
The term stenosis refers to a "narrowing." In the disease called mitral valve stenosis there is a narrowing of the opening from the left atrium into the left ventricle, usually due to a thickening of tissue in the mitral (bicuspid) valve itself. Mitral valve stenosis is one of the most common types of heart valve disorders, particularly in women. Because of the stenosis, the flow of blood into the left ventricle is inhibited; therefore, blood will often backflow into the left atrium thus reducing flow into the aorta.

Simulate stenosis by selecting Mitral Valve Stenosis from the popup menu and then clicking on the Apply Case button. Observe what happens to each of the cardiovascular parameters in this patient. Follow this patient for several seconds until you have a clear picture of which cardiovascular parameters may be changing.

> **What happens to blood pressure in this patient? Based on what you know about the neural control of heart rate, discuss what is happening to heart rate and explain why this is occurring.**

Assignment 9: Group Assignment
In the previous assignments, you examined many different aspects of the cardiovascular system to learn about the homeostasis of blood pressure. Work together in a group of four students to complete these exercises.

Divide your group into pairs. Have each pair create a cardiovascular condition where one parameter has been changed to create an abnormal situation. Run the experiment and show it to the other pair in your group. Ask this pair to identify the problem and suggest possible interventions, then have them test their possible interventions to see if they can bring about the desired effect.   Discuss the following questions.

Was the group successful in diagnosing the problem? Why or why not? Which interventions were suggested? Did the interventions work? Why or why not? Are the interventions that were suggested realistic manipulations that might be used in a clinical setting? Discuss your experiments and results with your instructor to help you answer this question.

DemographyLab

Background

Ecology includes a broad and fascinatingly complex range of topics related to the interrelationships between living organisms and their nonliving environments. Ecologists classify these interactions into a hierarchy based on the complexity of interrelationships that occur between living (**biotic**) components, and nonliving (**abiotic**) components (such as soil, water, weather, temperature, pollutants, light, and nutrients). At the simplest level of this hierarchy, organismal ecology focuses on the interactions between individual organisms and the environment. Individual organisms can be grouped into a **population**, which is defined as a group of individuals of the same species that occupy and are adapted to a particular geographic area at the same time. Populations of different species that inhabit a particular area at the same time are grouped together into **communities**. All of the organisms and nonliving components within a given area are classified as an **ecosystem**. At the highest level in this hierarchy, all ecosystems on Earth are part of the **biosphere**.

Population biologists are interested in the many factors that influence the size, health, growth, and distribution of a population. The scientific study of the factors that affect population size and growth over time is called **demography**. Two important factors that affect population growth are population **density** and the spacing or **dispersion** of individuals within the geological boundaries in which a population lives.

Ecologists have established a variety of methods for estimating the density of a population and for studying density-dependent and density-independent factors. Depending on the abundance and wariness of a species, sampling techniques can include signs of an organism such as nests, droppings, tracks, and reduction of specific forage.

Ecologists who study dispersion are particularly interested in the patterns of spacing that many species exhibit within a particular geographical range. Interactions of individuals within a population and the influence of abiotic factors will affect the spacing patterns of individuals within their geographical range. These patterns can involve grouped (clumped), evenly spaced (uniform), and unpredictable (random) dispersions of organisms.

Both density and dispersion of a population are directly influenced by three primary factors: (1) reproduction or birth rate (fertility), (2) death rate (mortality), and (3) migration. In migration, the addition of an organism to a population is known as immigration while the loss of an organism from a population is called emigration. Population growth remains the same (zero population growth; ZPG) if the number of individuals that join a population, either by immigration or by birth, is equal to the number of individuals that leave a population by either emigration or death. If the number of births and immigrating organisms exceeds the number of deaths and emigrating organisms, then population growth occurs. The reverse is true when a population is on the decline. In addition, both the birth rate and death rate of a population rely on the age of its members (**age structure**). For example, a population that has a higher percentage of older individuals that may be past their reproductive prime compared to younger individuals is likely to show a higher degree of mortality compared to birth rate.

Given the aforementioned factors and additional factors such as generation time and sex ratios, biologists, demographers, and sociologists can use statistical data and computer programs to predict future changes in the size of human populations. When studying populations of other organisms, factors such as competition and predator-prey relationships must also be considered.

Human population growth has wide-reaching impacts on human health and the infrastructure, resources, and carrying capacity of a given geographic region. The importance of understanding human population growth is exemplified by the following quote from the opening chapter of *The Population Explosion*, by Paul and Anne Ehrlich: "The population of the United States is increasing much more slowly than the world average, but it has more than doubled in only six decades; from 120 million in 1928 to 240 million in 1990. Such a huge population expansion within two or three generations can by itself account for a great many changes in the social and economic institutions of a society. It is also very frightening to those of us who spend our lives trying to keep track of the implications of the population explosion."

DemographyLab is designed to simulate how different demographic factors may influence human population size and growth, although most of the factors that you will consider in this lab have similar effects on the population growth of other animals and plants.

References
1. Kates, R. W. "Sustaining Life on the Earth." *Scientific American*, October 1994.

2. Raloff, J. "The Human Numbers Crunch." *Science News*, June 1996.

3. Ehrlich, P. R., and Ehrlich, A. H. *The Population Explosion*. New York: Simon & Schuster, 1990.

Introduction
DemographyLab will help you learn about the current demographics for human populations of seven countries. You can use this lab to investigate how population size, population structure, mortality rate, and fertility rate influence the size and growth of human populations over time. For simplicity, the effects of immigration and emigration are not included in DemographyLab. However, because the demographic data presented in this lab have been determined from actual current demographic statistics for each country, the population changes you will examine may simulate future demographic realities for these seven countries.

Objectives

The purpose of this laboratory is to:

- Demonstrate fundamental primary demographic factors that influence human population size and growth.
- Simulate the effects of demographic factors on human population growth.
- Use demographic models to predict future changes in human population growth.

Before You Begin: Prerequisites

Before beginning DemographyLab you should be familiar with the following concepts:

- The four major levels of study in ecology based on the complexity of interactions between organisms and their environment (see Campbell, N. A., Reece, J. B., and Mitchell, L. G., *Biology* 5/e, and Campbell, N. A., and Reece J. B., *Biology* 6/e, chapter 50).
- Important determinants of organism distribution within the biosphere (chapters 50 and 52).
- The definition of demography and examples of important factors that affect human population size and growth (chapter 52).

Assignments

For your ease in completing each assignment, the background text relevant to the experiment that you will perform is *italicized*, instructions for each assignment are indicated by plain text, and questions or activities that you will be asked to provide answers for are indicated by **bold text**.

The following assignment is designed to help you become familiar with the operation of DemographyLab.

Assignment 1: Getting to Know DemographyLab: Demographic Differences Among Nations

Countries differ with respect to population numbers, age structure, and fertility and mortality rates. These differences are caused by many factors, such as geographic size and location, level of economic development, government policies, and religious practices. The following exercises are designed to help you understand the influence of some of these factors on the size and growth of human populations.

1. Select the Population Structure view on the input screen of Demography Lab. Using the Country popup menu, examine the estimated 1998 population structure of each nation.

   Can you classify the population structures into two general patterns? Consider what you know about each of these countries. What do you think is the biggest factor distinguishing these two groups of nations?

2. Select the Fertility Rate view on the input screen of DemographyLab. Using the Country popup menu, examine the estimated 1998 fertility rates of each nation. (Try changing the scale to magnify these differences.)

Do you see any trends for the fertility rates compared to the population structures? What are they?

3. Select the Mortality Rate view on the input screen of DemographyLab. Using the Country popup menu, examine the estimated 1998 mortality rates of each nation.

   a. **Do you see any general similarities and differences between nations? Provide possible reasons for these similarities and differences.** *Hint*: To help you determine differences in mortality rates, click on the Run button and choose the Vital Rates view. Export the data from this table to your notebook by clicking on the Export Data button. Click OK to export data to your notebook as text.

   b. **Do you see any trends for the mortality rates compared to the age structures? What are they? Which age groups show the biggest differences in mortality rate among nations?**

   c. **Do you see any differences in the mortality rates of males compared with females for any of the countries? What are they? What might account for some of these differences?**

4. *Most scientists believe slower population growth is beneficial to the sustainability and quality of human life on Earth. In the mid-1980s, the World Health Organization (WHO) Task Force on Contraceptive Vaccines specifically designated a large sum of research funds for scientists to design novel contraceptive methods that could be used to slow population growth. The WHO identified certain developing nations of Africa, South America, and Asia, including India, as populations that could benefit from new contraceptive approaches (such as contraceptive vaccinations) that could easily be administered on a large scale.*

Select the Fertility Rate view on the input screen of Demography Lab. Examine the fertility rates for Nigeria and India.

   a. **Develop a list of differences and similarities that appear for the fertility rates of each country. Why might each country show these differences and similarities?**

   b. **Can you think of common societal issues shared by India and Nigeria that might account for the similarities?**

   c. **Provide possible explanations for why it may be difficult to establish contraceptive techniques in these countries that would have a significant impact on population growth.**

Assignment 2: Historical Effects on Demographic Changes

For a variety of reasons, populations sometimes undergo periods when there are large changes in reproduction or mortality. For example, desert plant populations may produce a large cohort of seedlings following heavy rains, or a fish population may experience a peak in reproduction due to a La Niña weather year. Human populations exhibit similar demographic changes. For example, the history of human civilization includes many well-documented events, such as plague (Black Death of medieval Europe) and famine (Irish Famine of 1845–1847), that produced pronounced demographic changes in human populations. The following exercises illustrate some examples.

1. *In the United States, a peak in reproduction occurred after World War II from the late 1940s through the early 1960s; this produced the "baby boom" generation.*

   Select the Population Structure view on the input screen of DemographyLab. Using the Country popup menu, examine the estimated 1998 population structure of the USA.

   Can you find the baby boomers?

   Click the Run button and choose the Population Structure view. (You may wish to use the Scale button to make the differences in age structure more obvious.) Use the arrow buttons at the top of the view to advance the population structure forward in time. Follow the baby boom generation as they age. Compare the population in 1998 with the projected population for 2028.

   What do you think the social consequences will be for these changes? Provide at least three examples of social consequences that will result from these changes.

2. *In 1980, China's government adopted a policy advocating one child per couple.*

   Select the Population Structure view on the input screen of DemographyLab. Using the country popup menu, examine the estimated 1998 population structure of China.

   Can you see evidence of China's population policy?

   Click the Run button and choose the Population Structure view. (*Hint*: You may wish to use the Scale button to make the differences in age structure more obvious.) Use the arrow buttons at the top of the view to advance the population structure forward in time by 5-year increments.

   What changes will occur in China's population structure as a result of its policy?

148

3. *At the same time the baby boom was occurring in the United States, there was a "baby bust" in post-World War II Japan, when fertility rates decreased.*

Select the Population Structure view on the input screen of DemographyLab. Using the Country popup menu, examine the estimated 1998 population structure of Japan.

Can you see evidence of the postwar baby bust? What has happened to fertility rates in Japan in recent years?

Click the Run button and choose the Population Structure view. (You may wish to use the Scale button to make the differences in age structure more obvious.) Use the arrow buttons at the top of the view to project the population structure forward in time.

If you were a demographer, what advice or concerns would you have for Japan's government in planning for the future?

Assignment 3: Stable Age Structure

If vital rates remain constant, what will happen to population structure over time? This question is investigated in the following exercises. Keep in mind that although we are working with a human population model, the same principles will apply for any population of plants or animals

1. Set the number of years to 300 and simulate the population growth of each of the seven nations by using the Population Structure view as described in the previous assignment. Examine the pyramid plots as they change every five years.

   What happens to each population over the long term? How do the plots differ for each nation? Which nations show the biggest changes over time? Speculate about why these changes occur.

2. *If a population's age structure stops changing from year to year, demographers say that it has a stable age structure. Two typical patterns of stable age structures are pyramids and inverted pyramids.*

   Compare the stable age structures and intrinsic growth rates of the seven countries. Do you see a pattern? Develop a hypothesis about the relationship between growth rate and the shape of the stable age structure. Test your hypothesis by selecting a nation and altering its fertility rates. Can you create the pyramid and inverted pyramid patterns?

3. *Suppose the human immunodeficiency virus (HIV), the virus that causes acquired immunodeficiency syndrome (AIDS), were to mutate into an airborne*

virus that could easily be transmitted by casual contact with infected individuals.

Choose any country, decrease its 1998 population size based on mortality due to this new strain of HIV, and simulate the population for 300 years. Note: Be sure to change mortality rates for both males and females.

What effect, if any, does this have on the long-term age structure? Examine the pyramid plots as they change every five years.

Repeat the procedure for alteration in the population structure by decreasing the population of sexually active males and females in the 20-year-old categories. Examine the pyramid plots as they change every five years.

Formulate a hypothesis to describe how changes (increases and decreases) in fertility rate will affect the long-term population structure. Do the same for changes in female mortality rates. Design and carry out simulations to test your hypotheses.

Assignment 4: Zero Population Growth

Zero population growth (ZPG) is of special interest to demographers and ecologists because population numbers will remain constant over time. The net reproductive rate (NRR) is a good indication of how close a population is to ZPG, because it represents the average number of daughters that a woman is expected to produce in her lifetime. ZPG is attained when NRR equals one; every woman is exactly replacing herself. (It is not necessary to consider males, from a demographic point of view, since only women bear children.) If NRR > 1, a population will increase and if NRR < 1, a population will decrease. A value of NRR = 1 corresponds to an intrinsic growth rate of zero. ZPG is investigated in the following exercises.

1. Simulate each country's population growth for 300 years and look at the time series plot of the population size. **Form a hypothesis about which countries are closest to ZPG (that is, NRR = 1). Compare the NRRs of the seven countries. How well did you do with your hypothesis?**

2. Choose a country with a high population growth rate (for example, Nigeria) and modify its fertility rates until you get as close as possible to ZPG.

   What types of changes to fertility rates were necessary to attain ZPG? Reset fertility rates back to their default values and change female mortality rates until you get as close as possible to ZPG.

   What types of changes to female mortality rates were necessary to attain ZPG? Do the same changes to male mortality rates achieve ZPG? Why or why not? Given the types of changes necessary to achieve ZPG, what realistic options are open to a nation to reach ZPG?

Assignment 5: Demographic Momentum

Even though a country may have fertility and mortality rates that give a zero or negative growth rate, population size may increase for many years before leveling off or decreasing. Demographers call this demographic momentum. The following exercises illustrate this phenomenon.

1. Simulate population growth in China for 100 years.

    How long does it take for the population size of China to begin decreasing? Does any other country exhibit a similar phenomenon?

2. **Choosing a country with a negative growth rate, try modifying its 1998 population structure so it will exhibit demographic momentum.**

3. **Propose a hypothesis about what causes demographic momentum. Test your hypothesis by altering the default values of any country and simulating it in DemographyLab. Design an experiment to illustrate negative demographic momentum (a decrease in population size followed by a long-term increase).**

Assignment 6: Dependency Ratios

From a social and economic point of view, human demographers are concerned about the proportion (ratio) of workers in a population to those who are economically dependent on these workers, either directly or through taxes such as Social Security. the following formula, called the dependency ratio, is used as a summary statistic.

Dep. Ratio = 100 [(no. children under 15) + (no. elderly over 65)] / [no. persons 15 – 64]

We look at the dependency ratio in the following exercises.

1. **Referring to the Tabular Data view of DemographyLab, compute the 1998 dependency ratio for each of the seven countries. Compare and contrast each nation. Which component, children under 15 or elderly over 65, is most important for determining the dependency ratio for different nations?**

2. **Compute the dependency ratio for the USA for the years 1998, 2003, 2008, 2038. What happens to the dependency ratio as the USA enters the next century? Which component, children under 15 or elderly over 65, is most important for determining the changes in the dependency ratio? What are possible social or economic implications of these changes?**

3. **What do you think will happen to the dependency ratio for a country over the long term if vital rates remain constant? Test your hypothesis using DemographyLab.**

Assignment 7: Sex Ratios and the Marriage Squeeze

The sex ratio for a population or age group is the ratio of the number of males to females. A related concept is what demographers call the marriage squeeze. On average, husbands tend to be a few years older than their wives, so the ratio of men in one age group to women in the next younger age group is an indication of whether or not there is a balance in the number of potential spouses for either sex. The following exercises illustrate these concepts.

1. Choose any nation and compute the 1998 sex ratio for each age group.

    How does the sex ratio change with age? At what age does the sex ratio become close to 1? Repeat the computation for a second country. Are the trends similar?

2. **Propose a hypothesis to explain differences in sex ratio with age. Test your hypothesis by altering the vital rates of a country and looking at the sex ratio of different age groups after 100 years of population change.**

3. Using the Tabular Data view for the USA for 1998, compute the ratio of men 20–24 to women 15–29 to women 20–24 and so on, until you reach the last age group.

    At what ages is the marriage squeeze most pronounced? Compute the USA 1998 sex ratios for each group. Would the marriage squeeze be less of a problem if, on average, spouses were of the same age?

4. Choose and country and modify the fertility rates so that population growth is a large positive value, a large negative value, and ZPG. For each case, simulate 100 years of population growth and examine the sex ratio and marriage squeeze at the end of the 100 years.

    Is there an effect of population growth rate on sex ratios? Is there an effect of population growth rate on the marriage squeeze? Can you explain these results?

Assignment 8: Exponential Population Growth and Decline

If vital rates remain constant, what will happen to population numbers over time? This question is investigated in the following exercises. Keep in mind that although we are working with a human population model, the same principles will apply for any population of plants or animals.

1. Set the number of years to 300 and simulate the population growth of each of the seven nations. Examine the Time Series graphs on both the linear and logarithmic scales.

What happens to each population over the long-term? These simulations assume that current fertility and mortality rates remain unchanged. Is this possible? Why or why not?

2. *Exponential growth is described by the following exponential equation:*

   $\underline{N}(t) = \underline{N}(0)exp^{rt}$ *equation (1)*

   where $\underline{N}(0)$ is the total population size at some arbitrary initial time, $\underline{N}(t)$ is the population size $\underline{t}$ years in the future, and $\underline{r}$ is the intrinsic growth rate of the population.

   What will happen if $\underline{r}$ is a positive value? What will happen if $\underline{r}$ is negative? How do the Time Series plots compare with the Growth Rate parameter listed in the summary Stats? Do the populations seem to follow this equation over the long term?

   If we take the logarithm of both sides of the exponential growth equation, we get:

   $log\ \underline{N}(t) = log\ \underline{N}(0) + \underline{rt}$ *equation (2)*

   What would you expect from a plot of the logarithm of population size over time? How the case for $\underline{r}$>0 differ from $\underline{r}$<0? Examine the Time Series plots of population size versus time for each of the seven nations with the Logarithmic Scale option selected. Do the logarithmic plots of the long-term population values agree with your predictions?

3. *Let's use equation (1) to predict population size. Choose any nation and simulate population growth for 300 years. Using the Intrinsic Growth Rate parameter, try predicting the 2298 population size using the population sizes for 1998, 2098, and 2198. (Set $\underline{r}$ equal to the intrinsic growth rate, set $\underline{N}(0)$ to the population size for 1998, 2098, or 2198, and use $\underline{t}$ = 100, 200, or 300 years.) Since the intrinsic rate is displayed to three significant figures, your predictions will be limited to three significant figures of precision.*

   How well does the exponential growth model do with the three predictions? Which predictions are most accurate? Can you explain why this might be the case? Try another nation to see if you get similar results.

4. *Another measure of population growth is the doubling time, if $\underline{r}$> 0 or half-life, if $\underline{r}$<0. The doubling time is the number of years it takes the population to double in numbers. The half-life is the amount of time it takes the population to decrease by 50%. The doubling time or half-life is given by the equation:*

   $\underline{T}=0.6931/|\underline{r}|$ *equation (3)* *where |r| refers to the absolute value of the intrinsic growth rate. If r > 0, T is a doubling time; if r < 0, T is a half-life.*

Compute the doubling times or half-lives for each of the seven nations. What are the political and social implications of these values?

5. Choose any country, alter its 1998 population size, and simulate the population for 300 years.

What effect, if any, does this have on the long-term population growth rates? Repeat the procedure for alterations in the population structure, fertility rates, and female and male mortality rates. Come up with a hypothesis for how changes in each of these factors affect the long-term population growth rates. Design and carry out simulations to test your hypotheses.

6. Try deriving equation (3) from equation (1). [*Hint*: Set $\underline{N}(t)$ equal to $2\underline{N}(0)$ or $\underline{N}(0)/2$.]

Assignment 9: Group Assignment

Given your observations on the differences in the demographic variables considered in the previous assignments, work in groups of four or five students to develop a hypothesis to consider which nations will exhibit positive long-term growth rates and which ones will have negative long-term growth rates. For each nation, click the Run button to simulate 100 years of population change and look at the Stats view of the output. The last three statistics are all measures of long-term population growth rates. How well do your rankings agree with the rankings of the actual measures of growth rate? Compare and contrast the summary statistics for the different nations.

Pick one of the nations that exhibits a positive long-term growth rate and discuss the following:

1. Should this nation take measures to limit its population size? If so what should those measures be? For example, should a nation limit immigration?

2. Should this nation use either voluntary or mandatory contraceptive programs to limit population size?

3. Identify environmental, health, and other quality-of-life issues that may arise as a result of long-term population growth.

4. Prepare a plan that could be used reduce population growth. If you were the leader of this country, how would you justify your plan to the citizens of your country?

Once you have discussed these topics within your group, compare your responses to those of other groups of students who evaluated the same country.

PopulationEcologyLab

Background

Even if you have never studied **ecology** before, it is likely that you have frequently observed important characteristics of population ecology in your daily life. For example, have you ever noticed an abundance of a particular species of animal such as birds, squirrels, geese, deer, or fish in certain locations around the area where you live while in other seemingly similar areas this animal appears nonexistent? Have you ever seen one species chasing, attacking, or feeding on another species? Understanding conditions that influence the abundance of organisms in a particular area and the interactions between both organisms of the same species and organisms of different species that share the same habitat is a primary interest of population ecologists.

For any **population**, a wide range of conditions can influence the number of individuals in the population and result in changes in population size and population growth. Some of these conditions include environmental changes such as weather, habitat quality, and food availability. Other conditions involve interactions between competing populations of the same or different species, and predator-prey relationships.

One reason population size is influenced by environmental factors is that most populations exist within defined boundaries of habitat. A population's boundaries may be vast, covering large areas of forest or ocean, yet the overall growth and health of the population's individuals is directly influenced by the geographical boundaries of the population. This is true in part because the boundaries of habitat in which a population resides define a finite **carrying capacity**—the maximum population size that a geographical area can support—dictated by available resources such as food and shelter. If the food supply of a geographical area is inadequate to support a population, organisms have several options. They may be forced to leave the area in search of a richer food supply, they may learn to obtain nourishment from another food source, or population size will diminish over time if the population is unable to adapt to changing conditions in resources. In some cases, exceeding carrying capacity can lead to extinction of a species.

In many **ecosystems**, competition for a common food supply from populations coexisting in an area can greatly influence the population size of all organisms relying on the common food source. Two species competing for the same limited resources ultimately cannot coexist. This concept is known as the **competitive exclusion principle**. In **interspecific competition**, populations of different species compete for resources within a community. However, competition between two populations is not an absolute requirement for depleting a food supply. Even if there is no competition for food from another population, the size of a single population of organisms may become too large, resulting in competition for a limited amount of food among individuals in the same population. Eventually this population may exceed its carrying capacity. This is an example of **intraspecific competition**.

Human actions and seasonal variations in weather can also wreak havoc on populations. Natural and manmade events that result in the destruction of habitat are frequently called **disturbances**. Disturbances such as fire, floods, damming rivers, habitat destruction due to construction, and drought—even if these disasters occur infrequently—can drastically affect population size by destroying habitat that a population relies on for necessities such as feeding, nesting, reproducing, resting, and hibernating. Natural disasters may greatly reduce population size and growth for some organisms while providing new opportunities for increased growth by other populations. As an example, a severe flood may kill populations of wild and domestic animals but the remains of these animals and the spread of nutrients over a wide area can provide new breeding habitats for bacteria and mosquitoes.

In addition to competition and disturbance, population growth can be restricted by biological events that exert strong effects on population size. Birth rate, the number of offspring produced by a population over a period of time, can affect populations both by increasing population size when birth rates are high and by decreasing population size when birth rates are low. Similarly, mortality rate—the number of deaths over a period of time—can greatly affect population size even when changes in mortality rate are small. Obviously, organisms that live longer (low mortality rate) have the potential to breed more frequently than organisms with a high mortality rate; thus, the potential for population growth is likely to be increased for a population with a low mortality rate compared to a population with a high mortality rate.

Even under conditions of adequate population size, adequate habitat and food supply, and ideal birth rates and mortality rates, many populations must deal with **predation**—an interaction that typically benefits the predator and is harmful to the population size of the prey. Predator-prey relationships can influence both the prey and the predator. Prey species are not always harmed by predation, nor do predator species always benefit from predation. For example, population size of prey can be controlled favorably through predation by preventing the prey population from exceeding carrying capacity. Population size for a predator species can be decreased when there is insufficient prey.

Population ecologists are often interested in studying patterns of population growth to prevent the likelihood of conditions that might cause a population to decline. As a result, population ecologists frequently use mathematical and computer modeling similar to PopEcoLab as a way of predicting how populations will be influenced in the future by current data they are collecting. For example, one may predict a model of exponential growth whereby a population may show an unlimited increase in size if resources are not limiting and mortality is not a problem. Through developing models, it is sometimes possible to predict when a population will exceed its carrying capacity and make choices about how to help prevent this from occurring. For example, in many areas of the northeastern United States, controlled harvests of white-tailed deer are frequently used to modulate population numbers in geographical areas that may be close to exceeding carrying capacity. For other animals, relocation of individuals to less populated areas is another strategy that can be used to maintain adequate population size.

You will use PopEcoLab to study important aspects of population ecology in two simulated populations of sparrows, brown sparrows and blue sparrows. In this hypothetical model, brown sparrows rely on seeds for their nutritional requirements and their carrying capacity is limited by seed density. Blue sparrows rely on insects to satisfy their nutritional requirements and their carrying capacity is limited by insect density. By manipulating seed density and insect density as important resources, the carrying capacity for each population of sparrows can be varied. These two sparrows are also potential competitors because brown sparrows can also eat insects and blue sparrows can also eat seeds. Hawks are potential predators on both types of sparrows. The ability of sparrows to avoid predation depends on their flight speed—at the highest flight speed, sparrows can avoid being killed by hawks.

You can also use PopEcoLab to control reproductive rates for all three species by varying clutch size, and mortality rates can be manipulated by varying the life span of each species. The experiments that you set up and analyze in PopEcoLab will provide you with an important understanding of the many intricate factors that influence population growth in two competing species and will enable you to study the effects of relationships in a predator-prey community.

References

1. Bush, M. B. *Ecology of a Changing Planet*, 2nd ed. Upper Saddle River, NJ: Prentice Hall, 2000.

2. Caswell, H. Predator-Mediated Coexistence: A Non-Equilibrium Model." *American Naturalist*, 112 (1978).

3. Rosenweig, M. L. Stability of Enriched Aquatic Ecosystems. *Science*, 175 (1972).

4. Smith, R. L., and Smith, T. M. Elements of Ecology, 4th ed. San Francisco: Benjamin/Cummings, 2000.

Introduction

This program simulates the dynamics of population growth for two populations of sparrows, and a population of hawks as predators of sparrows. In this laboratory, you will perform simulations of experiments to investigate a variety of topics in population ecology including the relationships between competing species, exponential growth, resource-limited growth, population extinction, population cycles, competitive exclusion, resource partitioning and coexistence, predator-prey cycles, and predator-mediated coexistence.

Objectives

The purpose of this laboratory is to:

- Demonstrate important relationships between the many factors that influence population growth of a species.
- Simulate changing conditions such as life span, clutch size, competition, food availability, and predator-prey relationships, and investigate the effects of these conditions on population size and growth.
- Investigate how populations can adapt to changing conditions that influence population growth.

Before You Begin: Prerequisites

Before beginning PopEcoLab you should be familiar with the following concepts:

- Important characteristics of population ecology including population density, population dispersion, and intraspecies[delete comma] and interspecies interactions (see Campbell, N. A., Reece, J. B., and Mitchell, L. G. *Biology*, 5/e, and Campbell, N. A., and Reece J. B., *Biology* 6/e, chapters 52 and 53).
- Factors that affect population growth such as limiting resources, birth rate, mortality, and sex ratios (chapter 52).
- How interactions between different species, and predation can result in changes in population density and population size (chapters 52 and 53).
- An understanding of how competition (competitive exclusion principle) can influence population size (chapter 53).

Assignments

For your ease in completing the following assignments, the background text relevant to the experiment that you will perform is *italicized*, instructions for each assignment are indicated by plain text, and questions or activities that you will be asked to provide answers for are indicated by **bold text**.

The following assignment is designed to help you become familiar with the operation of PopEcoLab.

Assignment 1: Getting to Know PopEcoLab: Single Species Population Growth

This assignment is designed to help you become familiar with the operation of PopEcoLab by studying the population growth of brown sparrows. The first screen that appears in PopEcoLab shows you an input parameter page with a table listing the default parameters for the laboratory conditions that you can manipulate when setting up your experiments.

Before you can set up any experiment in PopEcoLab, you must be familiar with the input parameters that you can manipulate. You will study the effects of these different parameters in future assignments. A brief description of each input parameter is provided below. Refer back to this section as needed when you are working on different assignments.

Click on the Change Inputs button to see all the parameters you can manipulate for this lab. A new page will open with buttons for each of the input parameters located at the left side of each page (initial population size will be open as the first input parameter). Click on each input parameter and read the descriptions below. Change each parameter so that you can become familiar with how each input parameter operates.

Initial Population – *the initial population for brown sparrows is set at 200 birds while the default value for blue sparrows and hawks is zero. Population size of each species can be manipulated by clicking and dragging the slider bar.*

Clutch Size – *clutch size is the number of eggs that a female bird lays in her nest. The default value for brown sparrows and blue sparrows is three eggs while the default clutch size for hawks is two eggs. Controlling clutch size is one way to influence the rate of reproduction for all three species of birds.*

Life Span – *brown and blue sparrows have a shorter life span than hawks. Notice that the default life span for sparrows is one year while the life span for hawks is three years. This feature allows you to manipulate mortality rates (independent of predation) to influence life span.*

Flight Speed – *the ability of sparrows to avoid predation by hawks depends on their flight speed. The default value for both brown and blue sparrows is 6 meters per second (m/sec).*

Competition – *although brown sparrows primarily eat seeds and blue sparrows primarily eat insects, the two sparrows are potential competitors because brown sparrows can also eat insects and blue sparrows can also eat seeds. The relative rates of consumption of these "alternative resources" can be varied independently.*

Resource Densities – *seed density and insect density directly influence the carrying capacity of brown and blue sparrows, respectively. The default values are 100 seeds per meter-squared (m^2) and 100 insects/m^2.*

After you have finished this introduction to the input parameters, click the Reset button at the left of the screen to return all input parameters to their default values.

1. Population Growth of Brown Sparrows: *Now that you are familiar with the basic parameters in PopEcoLab, set up the following experiment to help you understand some of the factors that affect population growth of a single species.*

   a. Leaving all input parameters at their default values, run a simulation of population growth by clicking the Run Experiment button. Note: The number of years to run each experiment can be manipulated in 100-year increments from 100 years to 500 years by using the popup menu in the lower left corner of the input parameter page. The default value is 100 years.

When the experiment has finished running, a separate page will appear that presents the results of your experiment. <u>Note</u>: Any of the following data views can be saved to disk or printed by clicking the Export button. Clicking on this button will open a separate window with your plot or table. From this window you can then save your data to your hard drive or a disk, and you can print your data by using the print feature of your browser software.

The following data can be examined:

<u>Population Size</u> – represented as a plot of the number of birds in the population versus years of the experiment.

<u>Phase Space</u> – a plot of the number of one species (e.g., blue sparrows) against another (e.g., brown sparrows) as population numbers change over time. This type of plot is valuable for examining relationships between species. You can select the species for each axis.

<u>Textual Data</u> – text columns of raw data for population size of brown sparrows, blue sparrows, and hawks.

<u>Input Summary</u> – a summary table of the input parameters for the experiment that you carried out.

Examine the population size plot. What is your estimate of the carrying capacity for the population of brown sparrows?

Repeat this same experiment at least three or four times to determine if the results from this experiment are consistent.

b. <u>Effects of Changes in Mean Clutch Size</u>
Formulate a hypothesis to predict the effects of a decrease in clutch size on the population number of the brown sparrow.

Prove or disprove this hypothesis by carrying out three different experiments with mean clutch size set to different values such as 0, 0.5, 1.0, 1.5, and so on, up to the maximum of 10 eggs, keeping all other parameters at their default values. Repeat each experiment several times, study the plots of population size, and then answer the following questions.

What did you discover? Did the results agree with your hypothesis? Why or why not? Is there a threshold clutch size needed to keep the population from going extinct? What happens to the variability in population numbers as clutch size gets smaller? If you were a

conservation biologist, what would you say about your ability to predict population numbers when clutch size decreases?

Formulate a hypothesis to predict the effects of an increase in clutch size on population number, then design and carry out experiments to test your hypothesis.

What happens to population size as clutch size gets larger? What happens to populations with very large clutch sizes? What do you think is causing the pattern that you see?

c. Mortality Rates: Investigate the effects of increasing mortality rates by decreasing life span and keeping all other parameters at their default values. Examine the plots of population size.

Is there a threshold life span needed to keep the population from going extinct? Explain this result.

Investigate the effects of decreasing mortality rates by increasing life span.

What happens to the variability in population numbers over time as life spans get longer? If you were a conservation biologist, what would you say about your ability to predict population numbers as life span changes?

d. Influence of Resource Density: Investigate the effects of decreasing and increasing seed density by running experiments with a decreased resource density for seeds and separate experiments with an increased resource density for seeds. For both experiments, keep all other parameters at their default values. Examine the plots of population size.

How does changing the amount of available resources affect population size? Why is this so?

2. Effect of Life Span on Minimum Viable Clutch Size (for use by advanced students): Formulate a hypothesis to explain how changes in life span might affect the threshold clutch size needed to keep a population of brown sparrows from going extinct, then design experiments to test your hypothesis.

What did you discover? Did the results of your experiment support or refute your hypothesis? Explain your results.

Assignment 2: Two Competing Species
As you learned in the background text for PopEcoLab, interactions between populations of two more different species (interspecific interactions) can affect population size and population density in a number of ways that can benefit or harm one or several species. These assignments are designed to help you learn about

interactions that result in competition by following the population growth of both the brown and blue sparrow.

1. Set the initial number of hawks to zero. Set the initial population numbers for both the brown and blue sparrow equal to 200 birds. Set the relative insect consumption by the brown sparrow equal to zero, and set the relative seed consumption by the blue sparrow equal to zero. Keep the remaining parameters at their default values. Run several simulations and get an estimate of the carrying capacity of each species. <u>Note</u>: You can run ten independent simulations by going to the Initial Population view and clicking the Multiple Run Mode button. With this mode, set the number of runs to 10, then run the experiment.

    What did you observe for the carrying capacity of each species?

    a. Go to the Competition view and use the slider to increase the relative insect consumption by the brown sparrow, then run this experiment.

        What happened to the density of the two sparrows? Is there a point at which one of the two species goes extinct?

        Now reverse the situation and keep the relative insect consumption by the brown sparrow equal to zero while increasing the relative seed consumption by blue sparrows. Before you run this experiment, formulate a hypothesis to predict the results of this experiment.

        What happened to the density of the two sparrows? Is that what you expected? Explain your results.

    b. Keep the two consumption rates of resources equal, but gradually increase both. Make sure you run several simulations for each set of consumption rates.

        What happens as you increase the amount of competition between the two species of sparrows? Explain these results.

    c. Set both of the relative consumption rates back to their default value of 0.50. Design experiments to investigate the effects of resource densities on species competition, then run these experiments multiple times. Summarize your results.

2. <u>Advanced assignment</u>: *An alternative way to view the type of ecological data that we have studied for these competition experiments is to plot the number of one species against the other species as their numbers change through time. This is called a "phase space" plot and the line that connects the data points is called a "trajectory" in phase space. If one or the other species goes extinct, the trajectory moves toward the x axis or y axis. If both species coexist, the trajectory moves toward a point on the interior of the phase plot.*

Try different combinations of competition between the two sparrow species. Set the number of generations to 500. On the Initial Population screen, click the Multiple Run Mode button and set the number of runs to 10. In the text fields, enter different combinations of initial numbers so that both species have low numbers, both have high numbers, the brown sparrow has low numbers while the blue sparrow has high numbers, and vice-versa. View results in the Phase Space view. On the phase space plot, set the Run Number to "All" to view all the trajectories together. Be sure to try examples where relative consumption rates are low, one is low and the other is high, and both are high.

Are there any situations where the initial sparrow numbers affect the outcome of competition? Explain why this may or may not be true.

Assignment 3: Predator and a Prey

Predation is a fact of life. Predators develop abilities to locate and capture prey, and prey can evolve to develop mechanisms designed to minimize predation. However, if predators are more effective at capturing prey than prey species are at avoiding predation, then the population of prey species will be affected. Conversely, for some species, too little predation can also affect population numbers in a negative way if the prey species population exceeds its carrying capacity.

1. Evading Predators - Set the initial number of blue sparrows to zero. We will focus on the population growth of the brown sparrow and the hawk. Set the initial population number for the brown sparrow equal to 200 birds and for the hawk equal to 20 birds. Design experiments to test the effects of increasing flight speed of the brown sparrow. Try several different flight speeds.

   What happens? How does increasing flight speed affect the population numbers of sparrows and hawks?

   Design experiments to test the effects of decreasing flight speed of the brown sparrow.

   What happens to the population numbers of the sparrow and hawk?

2. The Effect of Clutch Size on Predator-Prey Dynamics - Design an experiment to test the effects of changing sparrow clutch size on the predator-prey dynamics.

   What happens to the population numbers of the sparrow and hawk as sparrow clutch size increases? What happens when sparrow clutch size decreases?

   Design experiments to test the effects of changing hawk clutch size on the predator-prey dynamics.

What happens to the population numbers of the sparrow and hawk as hawk clutch size increases? What happens when hawk clutch size decreases?

3. <u>Paradox of Enrichment</u> - Rosenweig (1972) identified a phenomenon he called the "paradox of enrichment." He discovered that increasing the carrying capacity of a species—for example, by adding nutrients to a pond—can destabilize the system when the species is at risk of predation. Design experiments to test this idea.

   What did you discover? Explain your answers.

Assignment 4: Advanced Assignments: Two Competing Prey and a Predator

Ecologists have identified a phenomenon called "predator-mediated coexistence" (Caswell 1978), where a predator feeds on a dominant competitor, which allows the weaker competing species to coexist.

Design experiments where the brown sparrow always outcompetes the blue sparrow in the absence of any hawks, but by adding hawks that feed more heavily on the brown sparrow, all three species coexist.

Were you able to achieve these conditions? Explain your results.

Some predators are "specialists," which survive by focussing on one or a few abundant prey species. Others are "generalists," which need to eat a variety of less common prey to survive.

Design an experiment where a predator is unable to survive with only one prey source, but can persist when two prey species are available.

Were you able to achieve these conditions? Explain your results.

Ecological systems are complex. Sometimes one species can indirectly affect another through its effect on a third species.

Design an experiment where hawks eat only brown sparrows (i.e., blue sparrows have the maximum flight speed needed to avoid hawks) and hawks coexist with brown sparrows when blue sparrows are absent. However, when blue sparrows are present, their competition with brown sparrows is enough to drive hawks to extinction.

What did you discover? What parameters enabled you to create this situation?

Assignment 5: Group Assignment

The experiments that you have conducted so far were designed to help you learn about the many factors and interactions that can influence the size and growth of a population. Since ecosystems are affected by random events such as changes in weather, applied ecologists must often make an assessment of "extinction risk." They do this by constructing simulation models similar to PopEcoLab and running multiple simulations. This group assignment is designed to have you use PopEcoLab to evaluate extinction risk by running multiple simulations (e.g., 100 simulations) and noting the frequency with which a population becomes extinct.

Design experiments to study how this risk of extinction changes for a combination of parameters such as clutch size and life span.

Examine how the extinction risk varies for different time windows, for example, 100 years, 200 years, and so on. Be systematic in your experimental design. Devise a plan to divide your investigation among different members of your group and run multiple simulations for each experiment. Tip: If you go to the Initial Population screen and click on the Multiple Run Mode button, you can set the number of runs to 10. This will give you ten independent simulations that you can examine by using the spinners on the output screen.

Did the results of your experiments support your hypothesis? If not, reformulate a new hypothesis and carry out additional experiments. Explain how the experimental conditions you set up were responsible for your results. Once you think you have an explanation for your findings, design and carry out additional experiments to support or refute your thesis.

Discuss the results of your experiment with your instructor to clarify questions that you may have about interpreting the results of your experiment and understanding the effects of the experimental conditions you created.